Y0-BTD-383

Developments in Soil Science 4

NON-AGRICULTURAL APPLICATIONS OF SOIL SURVEYS

Further Titles in this Series

1. I. VALETON
BAUXITES

2. IAHR
FUNDAMENTALS OF TRANSPORT PHENOMENA IN POROUS MEDIA

3. F.E. ALLISON
SOIL ORGANIC MATTER AND ITS ROLE IN CROP PRODUCTION

Developments in Soil Science 4

NON-AGRICULTURAL APPLICATIONS OF SOIL SURVEYS

EDITED BY

R.W. SIMONSON

Soil Survey, Soil Conservation Service, U.S. Department of Agriculture, Washington, D.C. (U.S.A.)

Reprinted from Geoderma Vol.10 No.1/2

ELSEVIER SCIENTIFIC PUBLISHING COMPANY
Amsterdam London New York 1974

ELSEVIER SCIENTIFIC PUBLISHING COMPANY
335 JAN VAN GALENSTRAAT, P.O. BOX 211, AMSTERDAM, THE NETHERLANDS

AMERICAN ELSEVIER PUBLISHING COMPANY, INC.
52 VANDERBILT AVENUE, NEW YORK, NEW YORK 10017

LIBRARY OF CONGRESS CARD NUMBER: 73-90964

ISBN 0-444-41199-2

WITH 49 ILLUSTRATIONS AND 24 TABLES

PRINTED IN THE NETHERLANDS

CONTENTS

Preface

NON-AGRICULTURAL APPLICATIONS OF SOIL SURVEYS

Agricultural use of soils predates recorded history. The growing rather than gathering of food, started some 9,000 years ago, has been called "the agricultural revolution" by archaeologists. Profound changes in the usefulness of soils are again in progress, although not of a magnitude to be called a revolution. Soils are being used for a growing list of purposes in the industrial nations of the world.

Soils are construction materials for highways, foundations for houses, and vehicles for waste disposal, to name part of the widening spectrum of uses. Furthermore, more and more land is needed for home sites, roads, parks, and playgrounds for expanding populations. Possibilities of selecting poorly suited soils for these purposes are continually increasing. The cost of mistakes, both in money and unhappiness, is substantial. Furthermore, many mistakes can be avoided if the kinds, distribution, and usefulness of soils are known.

Determining the kinds, characteristics, distribution, and extent of soils is accomplished through soil surveys. Related investigations can determine the limitations of individual kinds of soils for a variety of uses, both agricultural and non-agricultural. Related investigations can also determine what must be done to use individual kinds of soils for different purposes.

Traditionally, the information obtained in soil surveys has been applied in agriculture. The earliest systematic soil surveys were made to improve agricultural practice about three-quarters of a century ago. In contrast, non-agricultural applications of soil survey data are a relatively recent development. Such applications were limited prior to the last two decades. Consequently, the sum total of experience is not large, even though the number and volume of the applications have burgeoned during the last decade.

For these reasons, reports of non-agricultural applications of soil surveys have been assembled for this issue of *Geoderma*. Examples illustrate the use of information about soils in regional planning, urban expansion, community planning, the design and construction of highways, disposal of solid wastes, and disposal of liquid wastes. Included as well are records of longtime use of soil survey information in suburban expansion and foundation engineering. A bit more exotic are applications of the information gathered through soil surveys to archaeological studies, to breaches in dikes, to claims for earthquake damage, and to investigations of sediment accumulation. These applications suggest possibilities beyond the more common ones.

The purposes of assembly and publication of these reports is to inform soil scientists more widely about what has been done with information on the nature of soils and to indicate further opportunities for use of existing knowledge to promote the public welfare.

ROY W. SIMONSON
Editor-in-chief

Geoderma, 10 (1973) 1 – 26

THE USE OF SOILS DATA IN REGIONAL PLANNING

KURT W. BAUER

Southeastern Wisconsin Regional Planning Commission, Waukesha, Wisc. (U.S.A.)

(Accepted for publication July 24, 1973)

ABSTRACT

Bauer, K.W., 1973. The use of soils data in regional planning. *Geoderma*, 10: 1 – 26.

The growing concentration of population in metropolitan centers and the concomitant diffusion of urban development across large areas of the earth's surface is a worldwide phenomenon. This phenomenon is creating areawide developmental and environmental problems of an unprecedented scale and complexity. The problems involved in providing economically feasible facilities for water supply, for sewage disposal, and for storm water drainage; in controlling pollution of streams and lakes, ground water, and air; in providing safe and rapid air and surface transportation; and in maintaining the overall quality of the environment within those large urban regions are, even when considered individually, among the most complex problems facing society. These problems are, moreover, all closely linked to far more basic problems of land and water use and are thereby inextricably interrelated. The formulation of sound solutions to these problems, therefore, requires comprehensive, areawide planning efforts which recognize the existence of a limited natural resource base to which both rural and urban development must be properly adjusted in order to ensure a pleasant and habitable, as well as efficient, environment for life.

The soil resources of an area are one of the most important elements of the natural resource base, influencing both rural and urban development. A need exists, therefore, in any comprehensive, areawide planning program to examine not only how land and soils are presently used but how the soils can best be used and managed. This requires an areawide soil suitability study which maps the geographic locations of the various kinds of soils; identifies their physical, chemical, and biological properties; and interprets these properties for land use and public facilities planning purposes. The resulting comprehensive knowledge of the character and suitability of the soils can be one of the most important tools through which an adjustment of areawide urban development to the underlying and sustaining natural resource base can be accomplished.

This paper describes the application of such a soil suitability study in an actual regional planning program. It describes how soils information has been and can be used in both graphic and numeric form in areawide and local land use planning, zoning, and land subdivision control; in tax assessment and development financing; and in the location and design of such public works as sanitary sewerage, storm water drainage, and transportation facilities. The importance of definitive soils information to every major step in the planning process from the formulation of goals and objectives and supporting planning standards; through plan synthesis, test, and evaluation; to plan implementation is illustrated; and the importance of the detailed operational soil survey to sound regional planning and to sound development decision-making documented.

INTRODUCTION

The growing concentration of population in metropolitan areas is a worldwide phenomenon. In the United States, the population is being concentrated in about 200 large metropolitan regions (U.S. Dept. of Commerce, Bureau of the Census, 1971). Yet, within these metropolitan regions, the population is being decentralized, with a concomitant diffusion of urban development across large areas of the earth's surface. This urban diffusion is, in turn, creating areawide developmental and environmental problems of an unprecedented scale and complexity.

The problems encountered in providing economically feasible facilities for importing, diverting, and transporting potable water, sewage, and storm drainage; in controlling pollution of streams and lakes, ground water, and air; in providing safe and rapid air and surface transportation; and in maintaining the overall quality of the environment within these large urban regions are, even when considered individually, among the most complex problems facing society. These problems are, moreover, all closely linked to far more basic problems of land and water use and are thereby inextricably interrelated. The formulation of sound solutions to these problems, therefore, requires a comprehensive, areawide approach, an approach which has been called by some metropolitan planning and by others regional planning. As used herein, these two terms are interchangeable and are defined as comprehensive planning for a geographic area larger than a county but smaller than a state, united by economic interests, geography, and common areawide developmental and environmental problems.

Such planning must recognize the existence of a limited natural resource base to which both rural and urban development must be properly adjusted in order to ensure a pleasant and habitable environment for life. Land and water resources are limited and subject to grave misuse through improper land use and supporting public works development, misuse which may not only needlessly increase the cost of such development, but misuse which can have severe, adverse environmental impacts. In such a regional planning effort then, the selection of desirable areawide development plans from among the practical alternatives available must be based in part upon a careful assessment of the effects of each particular alternative plan on the underlying and supporting natural resource base.

Such emphasis on the natural resource base is essential if better regional settlement patterns are to be evolved and irreparable damage to limited and increasingly precious land and water resources avoided. The natural resources of a region are vital elements to its economic development and to its ability to provide and sustain a safe, healthful, and pleasant environment for life. These natural resources not only condition, but are conditioned by, regional growth and urbanization. Such emphasis, however, requires the collection and analysis of a great deal more information about the natural resource base and its ability to sustain development than has been collected before in major regional planning efforts. Such information must include definitive data on topography; bedrock geology; mineral deposits; surface and ground water resources; coastal and inland flooding; microclimatology; woodlands and wetlands; fish and wildlife habitat; areas having scenic, historic, scientific, and recreational value; and on soils.

NEED FOR SOILS STUDIES

The soil resources of an area are one of the most important elements of the natural resource base, influencing both urban and rural development. Much that is of importance to mankind takes place in the soil; and the soil is, directly or indirectly, the foothold for much of the life on earth. It is the natural medium for the growth of plants; its properties and life serve to stabilize wastes and purify water; and it serves as a foundation for buildings, roads and all other man-made land-based structures. Soils, therefore, are a most important and valuable resource; and mounting pressures upon land are constantly making this resource more and more valuable. As one writer has urged: "This slight and superficial and inconstant covering of the earth should receive a measure of care which is rarely devoted to it" (Shaler, 1891).

The soil resource has been subject to grave abuse and misuse through improper land use development. Serious health, safety, and pollution problems have been created by failure to consider the capabilities and limitations of soils during the planning and design stages of rural or urban development projects. Such problems are usually very costly to correct and may create personal hardships out of all proportion to the relatively simple steps required to avoid them. Such problems include malfunctioning septic tank sewage disposal systems, surface and ground water pollution, flood damages, footing and foundation failures, soil erosion, and stream and lake sedimentation. Knowledge of the soil resource and its ability to sustain development not only help to avoid such problems but can also contribute to reducing rural and urban development costs. Such practices as the placement of streets and highways on unstable soils, the excavation of basements and utility trenches in shallow bedrock areas, the development of industrial and commercial sites on steep slopes, and the construction of underground utilities in high groundwater areas all result in additional construction and site preparation costs in order to overcome the limitations of the soils for the desired use. Such increased construction and site preparation costs may include the costs attendant to the removal of poor soils and their replacement with stable materials, the blasting of rock, extensive grading and terracing, and the use of tight sheathing and dewatering systems to control groundwater seepage. It has been estimated that urban development on soils poorly suited to such development may cost up to 63% more than on soils well suited to such development (Southeastern Wisconsin Regional Planning Commission, 1966a). Such practices may, in addition, result in later increased maintenance costs.

To help avoid further abuse and misuse of this important element of the natural resource base, a need exists in any comprehensive regional planning program to examine not only how land and soils are presently used but how they can best be used and managed. This requires definitive data about the geographic location of the various kinds of soils; about the physical, chemical, and biological properties of these soils; and about the capability of these soils to support various kinds of rural and urban land uses. Most importantly, it requires the use of such data to the greatest extent possible in guiding both rural and urban development.

For planning application, the necessary soils studies should be designed to permit preliminary assessment on a uniform, areawide basis of:

(1) The engineering properties of soils as an aid in the design of desirable spatial distribution patterns for residential, commercial, industrial, agricultural, and recreational land use development.

(2) The biological properties of soils, including both agricultural and nonagricultural soil–plant relationships and natural wildlife relationships, as an aid in the design of desirable spatial distribution patterns for permanent agricultural and recreational greenbelts and open spaces.

(3) The suitability and limitations of soils for specific engineering applications, such as private on-site sewage disposal facilities, agricultural and urban drainage systems, foundations for buildings and structures, and water storage reservoirs and embankments as an aid in the planning and design of specific development proposals and in the application of such land-use plan implementation devices as zoning, subdivision control, and official mapping ordinances.

(4) The engineering properties of soils as an aid in the selection of highway, railway, airport, pipeline, and other transportation facility location.

(5) The location of potential sources of sand, gravel, and other soil-related mineral resources.

Such an areawide soil capability study is not intended to, and does not, eliminate the need for on-site engineering foundation investigations or the laboratory testing of soils in connection with the final design and construction of specific engineering works. Such an areawide study is intended to provide the means of predicting the suitability of land areas for various land uses and public works facilities and thereby to permit, during the planning stages of development, the adjustment of regional settlement patterns, broadly considered, to one important element of the natural resource base.

REGIONAL PLANNING IN SOUTHEASTERN WISCONSIN

An example of a comprehensive regional planning program emphasizing a comprehensive assessment of the effects of alternative development patterns on the natural resource base is that of the Southeastern Wisconsin Regional Planning Commission. The Commission is the official planning and research agency for one of the large urbanizing regions of the United States and exists to serve and assist the local, state, and federal units of government in planning for the orderly, economic development of the seven-county Southeastern Wisconsin Region.

Regional planning as conceived by the Commission is not a substitute for, but a supplement to, federal, state, and local public and to private planning efforts. Its objective is to aid in the solution of areawide development problems which cannot be properly resolved within the framework of a single municipality or single county. As such, regional planning has the following three principal functions.

(1) Inventory; that is, the collection, analysis, and dissemination of basic planning and

engineering data on a continuing, uniform, areawide basis, so that the various agencies of government, private enterprise, and interested citizens within the region can better make decisions concerning community development.

(2) Plan design; that is, the preparation of a framework of long-range plans for the physical development of the region, these plans being limited to those functional elements having areawide significance. The permissible scope and content of the regional plan, as outlined in the state enabling legislation, extend to all phases of regional development, but implicitly emphasize the preparation of alternative spatial designs for the use of land and for the supporting transportation and public utility facilities.

(3) Plan implementation; that is, the promotion of regional plan implementation through the provision of a center for the coordination of the many planning and plan implementation activities carried out on a day-to-day basis by the various levels and agencies of government operating within the region.

The work of the Commission is visualized as a continuing planning process providing many outputs of use throughout the region, outputs of value to the making of development decisions by public and private agencies and to the preparation of plans and plan implementation programs at the local, state, and federal levels.

APPLICATION OF SOILS DATA IN REGIONAL PLANNING PROGRAM

The purpose of this paper is to illustrate how soils data have actually been utilized in the comprehensive planning program for the Southeastern Wisconsin Region. Such use can probably best be illustrated by relating the use to the regional planning agency's basic functions of inventory, plan design, and plan implementation.

Soils data and the inventory function

Reliable basic planning and engineering data collected on a uniform, areawide basis is absolutely essential to the formulation of workable development plans. Consequently, inventory becomes the first operational step in any planning process, growing out of program design. The crucial importance of factual information to the planning process should be evident since no intelligent forecasts can be made or alternative courses of action selected without knowledge of the current state of the system being planned.

The sound formulation of regional plans or major components thereof requires that factual data must be developed on: (1) the existing land use pattern; (2) the potential demand for each of the major land use categories and the major determinants of these demands; (3) the underlying natural resource base and the ability of this base to support land use development; (4) the existing and potential demand for transportation, utility, and public facility services and on the major determinants of these demands; and (5) the existing and potential supply of transportation, utility, and public facility system capacities.

The data must be developed through major planning inventories; and these inventories, when considered together, must be comprehensive, encompassing all the various factors which influence, and are influenced by, regional growth and development. Each inventory must be in a form which permits any individual finding to be related to the whole. The data collected in the necessary inventories must be pertinent to describing the existing situation with respect to regional development and identifying existing problems with respect thereto, forecasting future land use and public facility requirements, formulating alternative development plans, and testing and evaluating such alternative plans. It is important to note that in southeastern Wisconsin the framework for the necessary inventory efforts was provided by a regional systems analysis study which conceptualized the various mathematical models to be used in the areawide planning effort and then defined the data required to make these models operational.

One of the most important of the basic inventories carried out under the Commission work program concerned the soils of the region. At the time of the creation of the Southeastern Wisconsin Regional Planning Commission, a very limited amount of useful data on the soils of the region was available. General soil maps, which showed broad soil groupings at a small scale and provided limited interpretations for agricultural purposes, had been completed at various times in the past for each of the seven counties comprising the 2,689 square mile region. Modern standard soil surveys covering approximately one-third of the area of the region, but again accompanied by only agricultural interpretations, had also been completed in connection with the preparation of basic, individual farm conservation plans. In addition to being directed primarily at agricultural land use and treatment, these surveys had a further limitation in that they covered scattered small farms rather than broad areas.

The region lies within a wholly glaciated area, and this glacial history has created highly complex soil relationships and an extreme variability and intermingling of soils within even very small areas. The usefulness of generalized soils maps for definitive planning purposes within the region was, therefore, severely limited. The widespread occurrence of soils having questionable characteristics for certain types of urban development, coupled with the glacial history of the area, indicated the need for detailed soil surveys as an absolute prerequisite for sound areawide development planning. Moreover, adequate soil-suitability data were found necessary to the application of a regional land-use simulation model, a land-use design model, and a streamflow simulation model proposed to be developed by the Commission as integral parts of its planning program. The amount of land within each U.S. Public Land Survey section suitable for various types of urban land use, together with the hydrologic properties of the soils in each section, were necessary inputs to these models; and it was determined that these inputs could not be properly provided without detailed soils data. It was further determined that the required soils data would provide an important basis for regional plan implementation.

Soil survey procedures

Historically the study of soils within the region had been directed primarily to single-purpose applications, and little attention had been given to soil potentials on a comprehensive, areawide basis. Particularly, the study of soils had been historically related to use for agriculture and forestry, with little attention given to the ways in which soil properties might influence urban uses of land. It was believed, however, that standard soil surveys, such as those conducted by the Soil Conservation Service, U.S. Department of Agriculture, if accompanied by appropriate interpretations, could be adapted to meet the basic soils data needs of the comprehensive regional planning program (Soil Survey Staff, 1951). These surveys are made by carefully examining the soil in its natural state and delineating areas of similar soils on an aerial photograph. The areas so mapped are keyed to a national classification system (Simonson, 1962; Klingelhoets et al., 1968; Soil Survey Staff, 1973), in which all soils identified as belonging to a given series have, within defined limits, similar physical, chemical, and biological properties, these properties being determined by field and laboratory tests. This makes it possible to predict the behavior of the mapped soils, based upon past experience with similar soils, under any proposed land use.

These surveys have certain limitations particularly with respect to depth surveyed and with respect to the possible inclusion of soils with slightly different properties within mapped areas because of map scale, time, and cost limitations. Nevertheless. these surveys represented the best available source of areawide soils information. The surveys are carried out by experienced soil scientists and constitute a valuable basic scientific inventory which, if accompanied by the necessary interpretations, has multiple planning and engineering uses.

In order to fulfill the soils data requirements of the regional planning program, a cooperative cost-sharing agreement was negotiated for the completion of modern standard soil surveys of the entire region, together with the provision of interpretations for comprehensive planning purposes. Specifications governing the work were drawn by the Commission staff and were incorporated in the interagency agreement. These specifications called for some kinds of information not normally provided in soil surveys, and the work was accomplished in accordance with the salient provisions of those specifications.

Mapping and photography. Base maps and current vertical aerial photographs prepared by the Commission under its ongoing planning program, both uniformly covering the entire region, were used as the base maps for the soil survey in order to assure full compatibility with other Commission work. More specifically, all soils mapping was done on ratioed and rectified aerial photographs prepared from Commission negatives utilizing Commission base maps for horizontal control.

Operational soil survey. Standard soil surveys (detailed operational soil surveys) were completed for all those areas of the region not previously so surveyed (estimated at approximately one million acres), and such surveys were carried out in conformance with the latest standard operational procedures of the National Cooperative Soil Survey as set forth in the U.S. Department of Agriculture *Soil Survey Manual*. Boundaries of soil mapping units were identified on prints of aerial photographs, prepared in accordance with the Commission specifications, as already noted; and all mapped soil areas were identified by a suitable legend. Field mapping was actually accomplished, in accordance with the specifications, at a scale of 1″ = 1,320′ (1:15,840), each field sheet consisting of a ratioed and rectified vertical aerial photograph covering an area of six square miles (six U.S. Public Land Survey sections). All previously completed standard soils mapping was transferred to such current photography in order to provide uniform coverage of the entire region. The Commission was furnished, on a work progress basis, with reproducible half-tone negatives of the completed field sheets, prepared in accordance with the specifications on dimensionally stable cronar base material at a scale of 1″ = 2,000′ (1:24,000), to match the scale of the Commission planning base maps. These reproducible negatives were suitable for the preparation of inexpensive prints by diazo process and clearly showed the results of the soils mapping with delineations and identifying symbols so that the prints could be used for planning and engineering purposes on a work progress basis, not only by the regional planning staff but also by state and local governmental agencies and private investors.

Finished photo maps, at a scale of 1″ = 1,320′ (1:15,840), again utilizing negatives provided by the Commission, were prepared to accompany the standard published soil survey of the U.S. Department of Agriculture. Each such finished photo map covered an area of six square miles (six U.S. Public Land Survey sections). Key planimetric features, such as highways, railroads, streams, lakes, cemeteries, and major structures were identified on the finished photo maps as were the U.S. Public Land Survey township, range, and section lines. It should be pointed out that the mapping procedures used represented a marked departure from the then standard U.S. Soil Conservation Service practices, which prepared field sheets and final map sheets from controlled aerial mosaics rather than from ratioed and rectified aerial photographs, which used a different mapping scale, and which were not directly related to the U.S. Public Land Survey system.

Soils data interpretations. A published report was made available immediately upon completion of all the field mapping in the region (S.W.R.P.C., 1966b). It contained the information necessary to utilize the soil survey data in plan preparation and implementation at both the regional and local level. This information included definitive data on soil properties and interpretations of these properties for planning and engineering purposes. Thus, each kind of soil within the region was rated in terms of the inherent limitations for specific land uses and engineering applications. These ratings included presentation of the pertinent properties of each soil type relating to:

(1) Potential agricultural use, including soil capabilities for common cultivated crops, crop yield estimates, woodland suitability groups, and crop adaptation.

(2) Wildlife–soil relationships, including capability of the different kinds of soil to sustain various food plants and cover for birds and animals common to the region.

(3) Non-farm plant material–soil relationships, including suitability of different kinds of soil for lawns, golf courses, playgrounds, parks and open-space reservations.

(4) Soil–water relationships by kinds of soil, including identification of areas subject to flooding, stream overflow, ponding, seasonally high water table, and concentrated runoff.

(5) Soil properties influencing engineering uses, including depth to major soil horizons important in construction of engineering works, liquid limit, plastic limit, plasticity index, maximum dry density, optimum moisture content, mechanical analysis, AASHO* and Unified** classifications, percolation rate, bearing strength, shrink–swell ratio, pH, depth to water table, and estimated depth to bedrock if within approximately 20 ft. of the ground surface.

Interpretations of these properties of each soil type for planning purposes were also provided, including:

(1) Suitability ratings for potential intensive residential, extensive residential, commercial, industrial, transportational, natural and developed recreational, and agricultural land uses.

(2) Suitability ratings for septic tank disposal field, building foundation for low buildings, trafficability, surface stabilization, road and railway subgrade and earthwork uses.

(3) Suitability ratings for use as a source material for road base, backfill, sand or gravel, topsoil, and water reservoir embankments and linings.

(4) Rating with respect to flooding potential, watershed characteristics, susceptibility to erosion, and susceptibility to frost action.

(5) Suitability for wildlife habitat and habitat improvement, lawns, golf courses, playgrounds, and parks and related open areas requiring the maintenance of vegetation.

All of the data collected and all of the interpretations were summarized in tabular form suitable for ready use in planning and engineering analyses (see for examples Tables 1–5).

---

*The AASHO System is the most widely used soil classification system for highway engineering purposes. It identifies soils according to the qualities of texture and plasticity and groups them with respect to performance as highway subgrade materials. Originally devised in 1931 by the U.S. Bureau of Public Roads and revised by the Highway Research Board of the National Academy of Sciences in 1945, this system was thereafter adopted by the American Association of State Highway Officials. This classification system groups soils of the same load-carrying capacity into seven basic groups, A-1 through A-7. The best soils for highway subgrades are classified as A-1 and then in descending rank order to the poorest, which are classified as A-7. A wide range of load-carrying capacity exists within each soil group; and, therefore, the groups are subdivided into subgroups through the use of an index number ranking from zero for the best subgrade soils to 20 for the poorest. Increasing values of the index number reflect a reduction in load-carrying capacity and the combined effect of an increasing liquid limit and plasticity index and of the increasing percentages of coarse material.

**The Unified System of soil classification was developed for the U.S. Army, Corps of Engineers, during World War II and subsequently expanded in cooperation with the U.S. Department of the Interior, Bureau of Reclamation, for application to embankment and foundation construction, as well as to roadway and airfield construction. Like the AASHO System, the Unified System identifies soils according to the qualities of texture and plasticity and groups them with respect to performance as engineering construction materials. The following properties form the basis of the soil identification: the proportion of gravel, sand, and fines; the shape of the grain-size distribution curve; and the plasticity and compressibility characteristics of the soil. Each soil is given a descriptive name and a letter symbol. Three soil fractions are recognized: gravel, sand, and fines, the latter consisting of silt or clay; and the soils are divided into three major divisions: coarse-grained, fine-grained, and highly organic. The coarse-grained soils are further divided into gravel and sand, and each is in turn further subdivided into four groups. The fine-grained soils are divided into silt and clay, and each is further subdivided into three groups. The highly organic soils comprise one group.

Table 1. Selected measured, and estimated chemical and physical properties of soils of southeastern Wisconsin.

Soil number, type & horizon: Symbol	Depth, in.	Classification: USDA texture	Unified	AASHO	Mechanical analysis, % passing sieve†: No. 10, 2.0 mm	No. 200, 0.07 mm	Maximum dry density lb./cu.ft	Optimum moisture content	Permeability, in./hr.
217, Bono silty clay loam									
A	0-12	si. cl. l.	MH	A-7	100	100	107	18	0.2-0.8
B	12-36	si. cl.	CH	A-7	95	95			0.2-0.8
C	36+	si. cl.	CH	A-7	95	95	109	18	0.05-0.2
297, Morley silt loam									
A	0-12	si. l.	ML	A-4	100	85	98	25	0.8-2.5
B	12-36	si. cl.	CH	A-7	100	95			0.2-0.8
C	36+	si. cl. l.	CL	A-6	100	90	120	14	0.2-0.8
299, Blount silt loam									
A	0-10	si. l.	ML	A-4	100	85	98	25	0.8-2.5
B	10-36	si. cl.	CH	A-7	100	100			0.2-0.8
C	36+	si. cl. l.	CL	A-6	100	95	120	14	0.2-0.8

Symbol	Depth, in.	Percolation rate, min/in	Liquid limit	Plasticity index	Shrink-swell potential	Bearing strength	Reaction (pH)	Depth to water table, feet	Susceptibility to erosion
A	0-12	120-300	55	30	High	Poor-low	6.6-7.3	‡	A-slight
B	12-36	120-300			High	bearing	6.6-8.5		
C	36+	300+	38	22	High	capacity	7.4-8.5	0-1	
A	0-12	31- 60	56	32	L-M§	Poor-low	5.6-7.0	3-5	A-slight
B	12-36	61-120			High	bearing	5.6-7.0	>5	BMCN-mod.
C	36+	61-120	27	12	Mod.	capacity when wet	7.4-8.5		DKEF-severe
A	0-10	31- 60	56	32	L-M§	Poor-low	5.6-7.3	0-2	A-slight
B	10-36	61-120			High	bearing	5.6-7.3		B-mod.
C	36+	61-120	27	12	Mod.	capacity when wet	7.4-8.5	3-5	

* All three soils are moderately susceptible to frost action, have low bearing strength, and a depth of more than 5 feet to bedrock. Variations in these properties would be shown in additional columns of the table. † 100% of all samples passed No. 4 sieve (4.70mm). ‡ Seasonal. § L - Low, M Moderate.

In the application of soils data, the planner is concerned not only with the properties of the various soils and the interpretation of these properties in terms of suitability for various land uses, but also with the spatial distribution of the various soils, their areal extent, and their accurate location with respect to other factors influencing regional development, such as existing land uses, utility service areas, accessibility patterns, and transportation service levels and areas. Moreover, the necessary soils information may be used either graphically, as for example, to show how soils having various properties are distributed relative to each other, to other elements of the resource base, and to existing land uses, or quantitatively, as for example, to determine the total area covered by soils having certain properties. Therefore, the soils data resulting from the detailed operational

Table 2. Suitability rating* of soils of southeastern Wisconsin for rural and urban land use development.

Soil type and number	For agricultural use	Residential development: With public sewer	Without public sewer: Less than one acre	Without public sewer: One acre or more	Commercial and industrial development	Transportation systems
76 Will Loam	Fair Good when drained (Poor for trees)	Poor	Very poor	Very poor	Good when drained	Poor
217 Bono Silty Clay Loam	Good when drained (Poor for trees)	Very poor	Very poor	Very poor	Poor	Poor
297 Morley Silt Loam	Good on 0-6% slope, Fair on 7-12% slope (Fair for trees)	Good	Question-able	Poor	Fair on 0-6 % slopes, Poor on slopes over 6%	Poor
298 Ashkum Silty Clay Loam	Good for crops when drained, Good for pasture, (Poor for trees)	Very poor	Very poor	Very poor	Poor	Poor
299 Blount Silt Loam	Good for crops when drained, Good for pasture (Fair for trees)	Fair	Very poor	Question-able	Poor	Poor
398 Ashkum Silt Loam	Good for crops when drained, Good for pasture (Poor for trees)	Poor	Very poor	Very poor	Poor	Poor

* Suitability rates apply to entire soil profile and its position in the landscape.

Table 3. Suitability ratings and limitations of soils of southeastern Wisconsin for specific engineering purposes.

Soil type and number	Limitations for: Road subgrade	Limitations for: Foundation for low buildings	Suitability as a source of: Topsoil	Suitability as a source of: Sand & gravel	Corrosion potential: Metal	Corrosion potential: Concrete
217 Bono Silty clay	Very poor	Very poor	Good-surface, Very poor-subsoil	Very poor	Very high	Low
297 Morley Silt Loam	Poor	Fair	Good-surface, Poor-subsoil	Very poor	Mod-erate	Low
299 Blount Silt Loam	Very poor	Poor	Good-surface, Poor-subsoil	Very poor	Very high	Low

Table 4. Suitability rating* of soils of southeastern Wisconsin for recreational development.

Soil type and number	Play-grounds, parks, and picnic areas	Bridle paths and nature and hiking trails	Golf courses	Cottages and utility buildings	Camp sites	Remarks
76 Will Loam	Poor	Poor	Poor	Very poor	Very poor	Suited to wildlife and ponds, High water table
217 Bono Silty Clay Loam	Poor	Poor	Poor	Very poor	Very poor	Suited to wildlife and ponds
297 Morley Silt Loam	Good on 0-2% slopes, Fair on 3-6% slopes	Good on 0-12% slopes, Fair on slopes over 12%	Good on 0-6% slopes, Fair on 7-12% slopes	Poor	Good on 0-6% slopes, Fair on 7-12% slopes, Poor on slopes >12%	
298 Ashkum Silty Clay Loam	Poor	Poor	Very poor	Very poor	Very poor	Suited to wildlife and ponds
299 Blount Silt Loam	Fair	Fair	Fair	Poor	Poor	Seasonally high water table, Subject to ponding
398 Ashkum Silt Loam	Poor	Poor	Poor	Very poor	Very poor	High water table, Suited to wildlife and ponds

* Suitability rating applies to entire profile including subsoil and substratum; suitability rating for all other use considers entire soil profile.

soil surveys were adapted for ready use in the regional planning program. This required appropriate transformation of the soils maps and interpretive data into the two forms in which it was actually used: graphic and numeric.

In order to permit the efficient application of the soils data in graphic form, the Commission staff prepared interpretive soils maps. These maps were prepared at a scale of 1″ = 2,000′ (1:24,000) by county as overlays to the Commission base maps and were reduced for publication at a scale of 1″ = 4,000′ (1:48,000). The interpretive maps were prepared for seven kinds of potential land use: (1) agricultural; (2) large lot residential without public sanitary sewer service; (3) small lot residential without public sanitary sewer service; (4) residential with public sanitary sewer service; (5) industrial; (6) transpor-

Table 5. Suitability ratings, limitations, and selected properties of the soils of southeastern Wisconsin for watershed management purposes.

Available water capacity, in./in.	Flooding potential	Irrigation potential	Suitability for reservoir embankments and linings
0-10 - .2 10-36 - .18 36+ - .16	Subject to ponding	Poor	Good to fair impervious Medium to low stability & high volume change
0-10 - .2 10-36 - .18 36+ - .16	None	Fair-A, B & M slope Poor- C & N slope Very poor- D, E & F slope	Good to fair impervious Low stability & large volume change
0-10 - .2 10-36 - .18 36+ - .16	Subject to occasional ponding or flooding	Fair at A & B slope	Good to fair impervious Low stability & large volume change

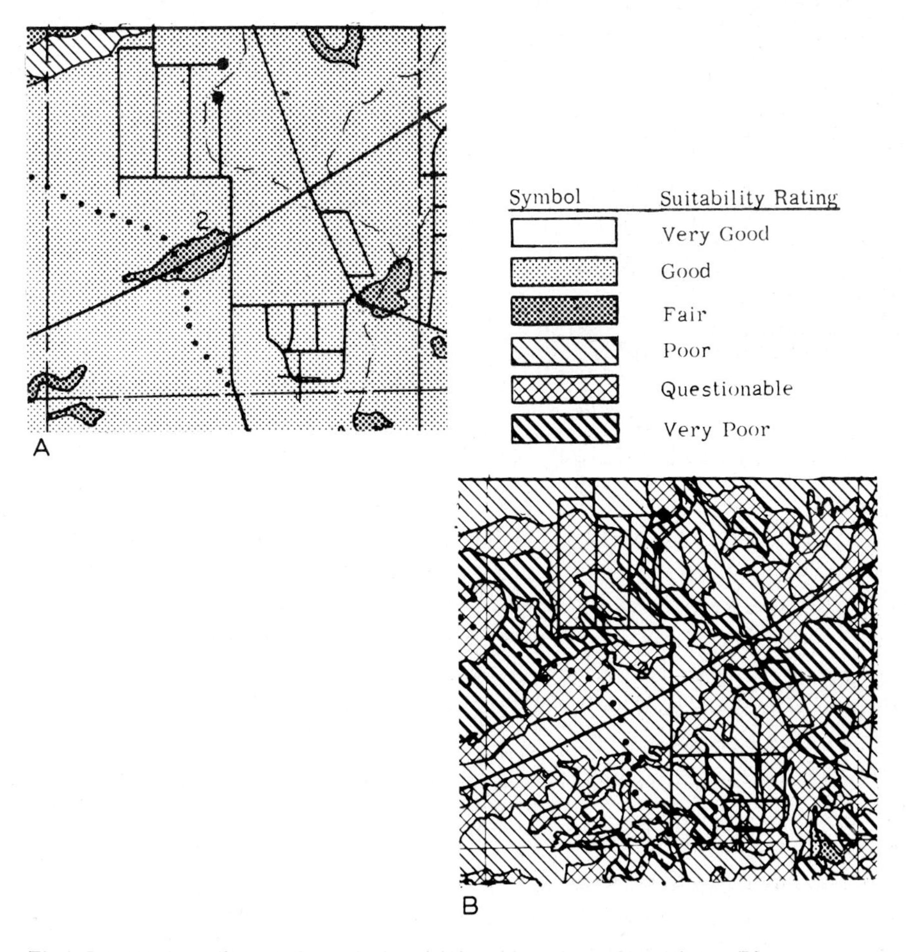

Fig.1. Interpretive soils maps for agricultural (A) and large lot residential uses (B).

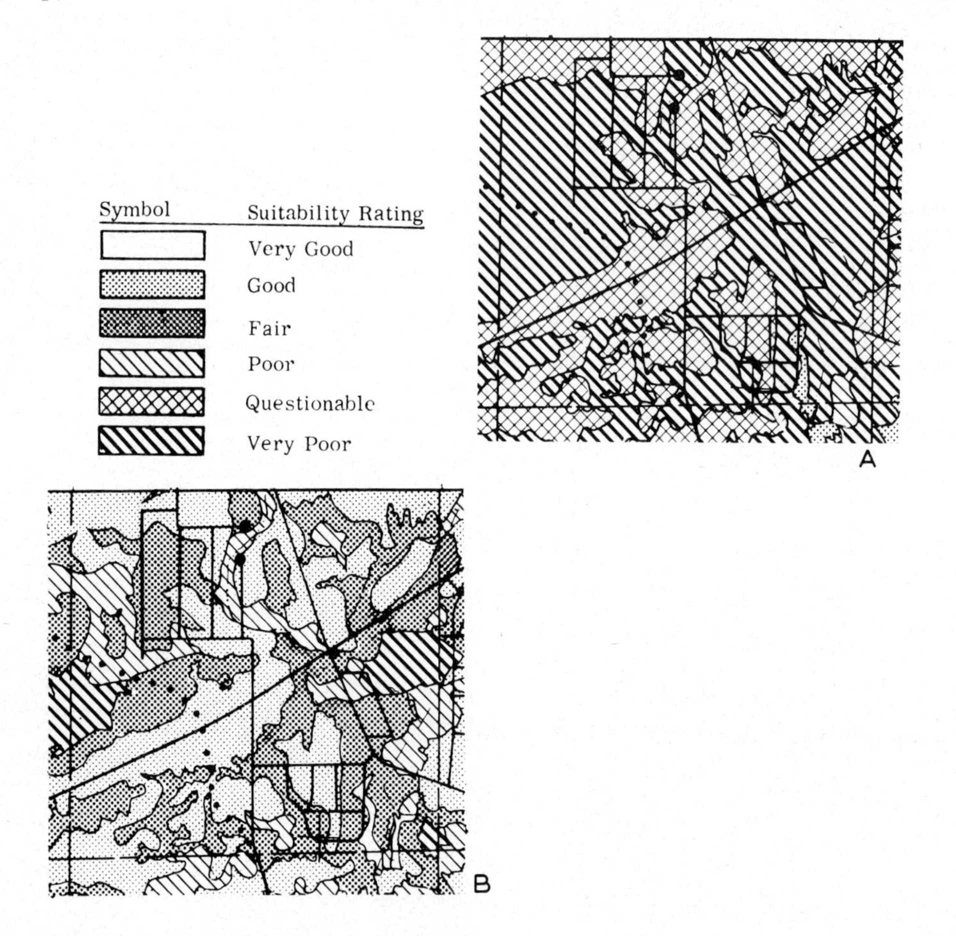

Fig.2. Interpretive soils maps for small lot residential use without public sewer service (A) and for residential use with public sewer service (B).

tation route location; and (7) intensely developed recreational. Each interpretive map will show the six soil limitation ratings: very slight, slight, moderate, severe, severe to very severe, and very severe (see Fig.1–4). These terms are defined as follows:

(1) Very slight: little or no soil limitations, very good suitability for use.

(2) Slight: slight soil limitations, easy to overcome during development, good suitability.

(3) Moderate: moderate soil limitations can be overcome with careful design and good management, fair suitability.

(4) Severe: severe soil limitations, difficult to overcome during development, poor suitability.

(5) Severe to very severe: very severe soil limitations, very difficult to overcome, require detailed on-site investigations, questionable suitability.

(6) Very severe: very severe soil limitations that lead to serious construction and maintenance problems, very poor suitability for use.

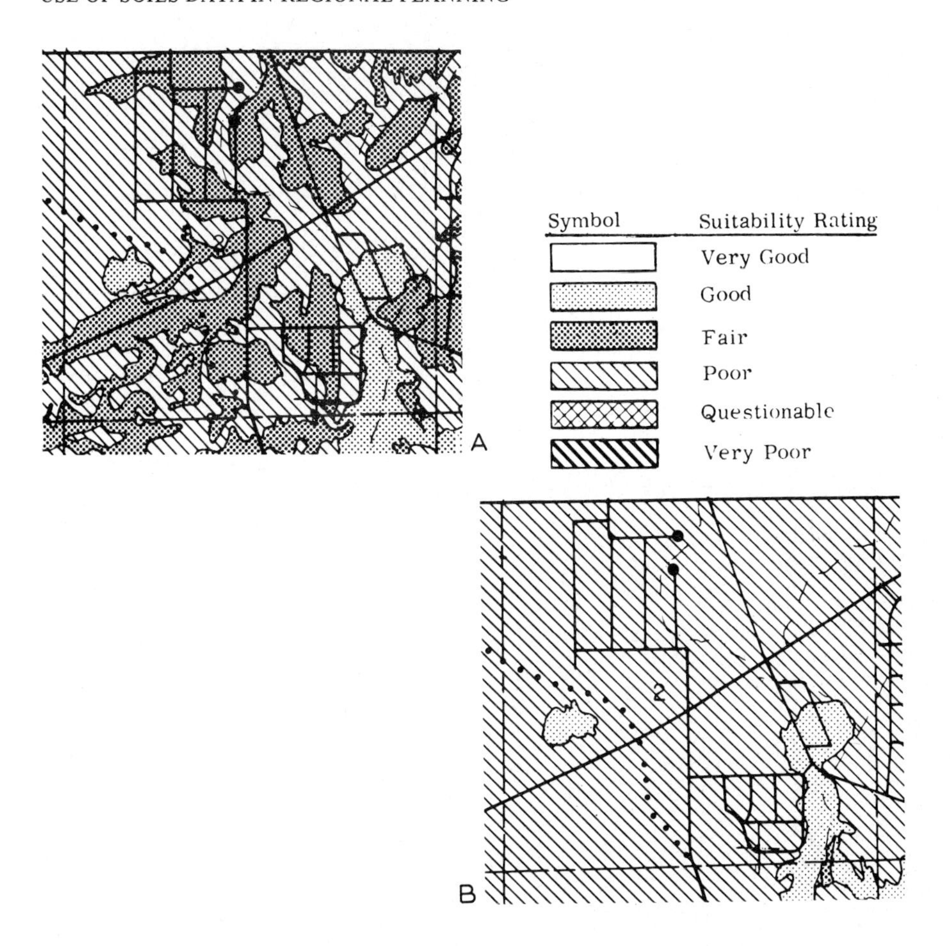

Fig.3. Interpretive soils maps for industrial (A) and transportation route uses (B).

In addition, a slope map was prepared for each county using the following slope ranges: 0 through 1%, over 1 through 5%, over 5 through 8%, over 8 through 11%, over 11 through 14%, over 14 through 19%, over 19 through 29%, and over 29%. Other interpretive maps for planning and engineering purposes based upon quantitative soil properties readily suggested themselves, and were prepared, as the regional planning work progressed to more definitive phases.

Quantification of mapped data

In order to permit the efficient application of the soils data in numeric form, the soil mapping units were measured by U.S. Public Land Survey section using random sampling techniques, and the percentage and total areas of each of the various soils within each section so determined, coded and transferred to magnetic tape for machine processing, tabulation, and use in planning analyses and model application. Once the basic data were

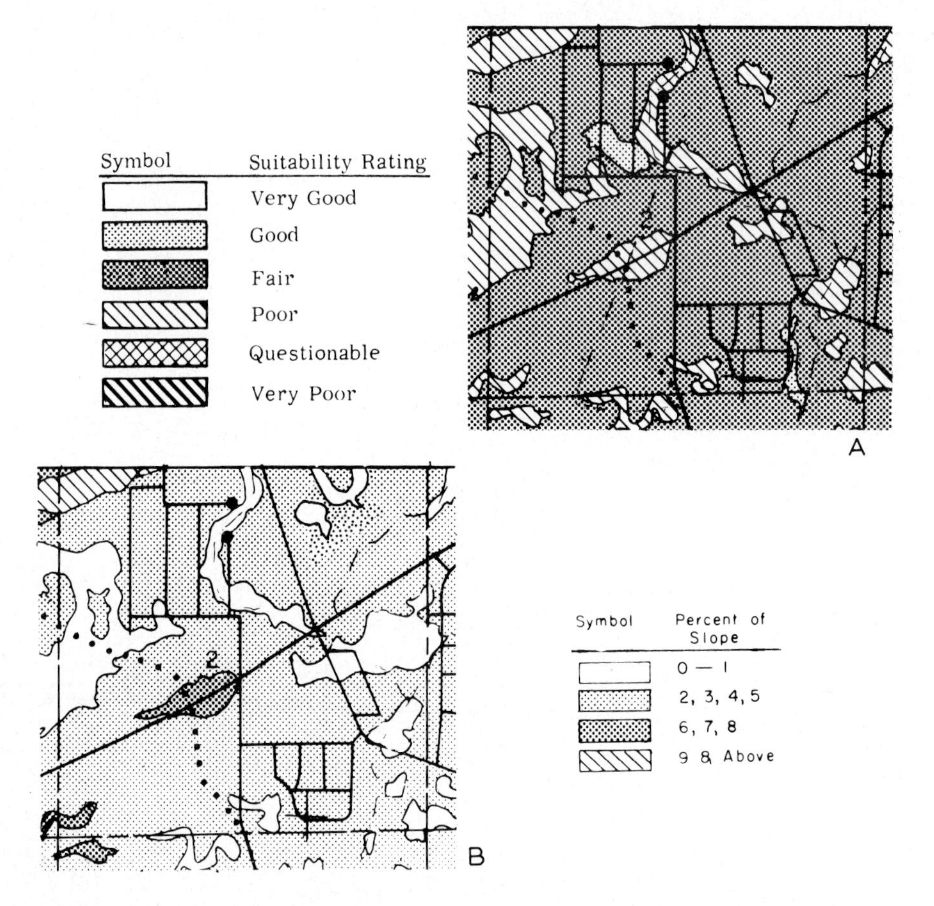

Fig.4. Interpretive soils map for intensely developed recreational use (A) and slope map for engineering purposes (B).

on magnetic tape, the percentage and total areas of each of the various suitability groupings within the region and any subareas of the region, such as watersheds, planning districts, or traffic analysis zones, were determined by machine methods. Even more important, however, is the fact that once the soils data were on magnetic tape they were readily correlated with other factors influencing development, the data for which could also be stored on magnetic tape. It should always be remembered that soils are only one of the factors influencing development decisions. Therefore, this correlation of soils data with other essential planning data by machine methods is an extremely important and useful advantage.

Use of soils data in plan design

One of the most important reasons for undertaking the regional soil survey in southeastern Wisconsin was to provide data essential to the preparation of regional land use and supporting public works facilities plans. Since the process of plan design is essentially a problem in finding the least costly way to meet stated objectives, in any planning effort it is necessary to link geographic location with development costs. In this way, alternative plans can be explored and the least costly alternative which meets the agreed-upon objectives adopted. Detailed soil surveys provide the means for relating development costs to geographic location since development costs vary with soil type and since the soil types have been geographically mapped. Accordingly, the regional soil survey was utilized in the comprehensive planning program for the Southeastern Wisconsin Region in the preparation of planning objectives and supporting standards, in the formulation of a regional land use plan, in the formulation of water resource management plans, and in the formulation of related public works facilities plans.

Planning objectives and standards

Planning may be defined as a rational process for formulating and meeting objectives. Before plans can be prepared, therefore, objectives must be formulated. In the regional planning program for southeastern Wisconsin, this task was initially undertaken as part of a regional land use planning program. Subsequent regional or subregional planning programs, such as the series of comprehensive watershed studies, have refined and extended the objectives initially formulated, as appropriate, to additional and more specific subject areas. Objectives are defined as goals or ends toward the attainment of which plans and policies are directed. In turn, standards are defined as criteria used as a basis of comparison to determine the adequacy of plan proposals to attain the stated objectives.

Several objectives and standards formulated and adopted by the Commission in its land-use planning program relate directly to the use of the regional soil survey and its interpretive analyses. In addition to the general objective of protecting, wisely using, and soundly developing the natural resource base of the region, the Commission has adopted the following specific development standards that are based upon the regional soil survey and its interpretive analyses.

(1) Urban development, particularly for residential use, shall be located only in those areas which do not contain significant concentrations of soils rated in the regional detailed operational soil survey as having severe or very severe limitations for such development. Significant concentrations are defined as follows: (a) in areas to be developed for low-density residential use, no more than 2.5% of the gross area should be covered by soils rated in the regional soil survey as having severe or very severe limitations for such development; (b) in areas to be developed for medium-density residential use, no more than 3.5% of the gross area should be covered by soils rated in the regional soil survey as having severe or very severe limitations for such development; (c) in areas to be developed for high-density residential use, no more than 5.0% of the gross area should be covered by soils rated in the regional soil survey as having severe or very severe limitations for such development.★

★These standards are based upon development of neighborhood units utilizing conventional land subdivision design layouts, with lot sizes throughout the neighborhood unit uniformly approximating

(2) Rural development, principally agricultural land uses, shall be allocated primarily to those areas covered by soils rated in the regional soil survey as having only moderate, slight, or very slight limitations for such uses.

(3) Land developed or proposed to be developed for urban uses without public sanitary sewer service should be located only in areas covered by soils rated in the regional soil survey as having moderate, slight, or very slight limitations for such development.

(4) New industrial development should be located in planned industrial districts in areas which contain soils rated in the regional soil survey as having only moderate, slight, or very slight limitations for such development.

(5) New regional commercial development, which would include activities primarily associated with the sale of shopper's goods, should be concentrated in regional commercial centers in areas which contain soils rated in the regional soil survey as having only moderate, slight, or very slight limitations for such development.

(6) All prime agricultural areas, defined as those areas which contain soils rated in the regional soil survey as having only slight or very slight limitations for agricultural uses and which occur in concentrated areas over five square miles in extent that have been designated as exceptionally good for agricultural production by agricultural specialists, should be preserved.

(7) All agricultural lands surrounding adjacent high-value scientific, educational, or recreational resources and covered by soils rated in the regional soil survey as having moderate, slight, or very slight limitations for agricultural use should be preserved.

(8) All agricultural areas which are covered by soils having moderate limitations for agricultural uses should be preserved in agricultural uses if these soils occur in concentrations greater than five square miles and surround, or lie adjacent to, areas which qualify as prime agricultural areas or occur in areas which may be designated as desirable open spaces for shaping urban development.

In its comprehensive water-resource management planning programs, the Commission has also adopted objectives and standards that relate to the regional soil survey and its interpretive analyses. To achieve the general objectives of reducing storm-water runoff, soil erosion, and stream sedimentation and pollution, the following standards have been formulated.

(1) A minimum of 50% of the area of the watershed in agricultural use should be under district cooperative soil and water conservation agreements and planned conservation treatment.

(footnote p.17 continued)
the average lot size required to meet the desired neighborhood population level and gross population density. If larger areas of a potential neighborhood unit than those specified above are covered by poor soils and are placed in open-space use without varying the lot size and subdivision layout, the population level and gross population density of the neighborhood unit may be adversely affected, as may the quality of the urban services provided. If variations in the subdivision layout design and lot size are permitted, such as cluster subdivision, minimum population levels necessary to sustain a desirable level of urban services may be achieved in areas covered by much higher percentages of poor soils than recommended in the standards, up to 75% of low-density neighborhoods, up to 50% of medium-density neighborhoods, and up to 44% of high-density neighborhoods.

(2) A minimum of 25% of the area of the watershed in agricultural use should be under conservation treatment.

To achieve the general objective of ensuring certain specified stream and lake water quality standards, the following standard has been formulated: all urban residential development, except single-family residences on lots of 5 acres or more in area and located on soils rated in the regional soil survey as suitable for the soil absorption method of sewage disposal, shall be served by centralized public sanitary sewerage facilities conveying liquid wastes to a sewage treatment plant.

The foregoing examples demonstrate the incorporation of soils data and interpretations directly into statements of regional planning development objectives and standards. Once stated, these objectives and standards become the guidelines for plan design, test, and evaluation.

Regional land use plan

The soil-survey data provided a particularly important input to the preparation and design of the adopted regional land use plan. The use of detailed soils data in such a large-scale regional land use planning effort was unprecedented. Four alternative regional land use plans were prepared – a controlled existing trend plan, a satellite city plan, a corridor plan, and an uncontrolled existing trend plan. The controlled existing trend plan was recommended for adoption and, after public hearings, was refined and ultimately adopted.

Plan design methodology

Interpretive soil maps at a scale of 1″ = 2,000′ were prepared using a "stop–go" color-coding system for those areas of the region that were as yet undeveloped for urban purposes but that had potential for such development. These interpretive soil maps were based on the interpretive ratings accompanying the detailed soil-survey maps. The particular interpretations chosen for this application included the interpretations for residential development served by public sanitary sewer systems, for residential development without public sanitary sewer on lots less than one acre in area, for residential development without public sanitary sewer on lots one acre or more in area, for commercial and industrial development, for agricultural use, and for transportation system development.

Once the interpretive soil maps were prepared, it was possible to measure and thus quantify the amount of land in each U.S. Public Land Survey section that had severe and very severe limitations for urban development even if served by public sanitary sewer. By subtracting this poorly suited area from the gross area of the section and by further subtracting areas committed to existing urban development, primary environmental corridor (less any poor soils in such corridor), and water, it was possible to arrive at a "net" land area for each section. This "net" land area was termed "developable land" and was assumed to be potentially available for future urban development. Once this process was completed, the alternative regional land-use plans were prepared using well-developed land-use planning techniques for balancing the forecast demand for, and supply of, land for the various uses and for spatially distributing the various needed land uses within the planning area.

It should be noted that the poorly suited soils, as defined above, were also important inputs to the delineation of the environmental corridors incorporated into the regional land-use plan. These corridors are defined as elongated areas encompassing the best remaining elements of the natural resource base, including, in addition to soils ill-suited for urban development, all major bodies of surface water and their associated floodlands; wetlands; woodlands; wildlife habitat areas; rough topography; significant geological formations; and several other features related to the natural resource base, including existing and potential outdoor recreation and related open-space sites, historic sites and structures, and significant scenic areas or vistas.

The regional land-use plan finally selected from among the alternatives available for adoption represents a conscious continuation of historic development trends within the region. Urban development would, in general, continue to occur in roughly concentric rings along the full periphery of, and outward from, existing urban centers. The plan proposes, however, to regulate, in the public interest, the urban land market in order to provide for a more orderly and economical regional development pattern, thus avoiding the intensification of areawide developmental and environmental problems.

The adopted regional land-use plan provides for the conversion of more than 71,000 acres of vacant and agricultural lands to residential use in the 27-year period from 1963 to 1990. This new residential development would take place in three density categories – low, medium, and high. Because so much of the urbanizing portion of the region consists of soils that have severe and very severe limitations for the proper operation of on-site soil absorption sewage disposal systems, the adopted plan proposed to serve all of the new medium- and high-density residential development with public sanitary sewerage facilities. This would mean that by 1990 over 95% of the total urban area within the region would be served by public sanitary sewerage facilities. All new low-density residential development which could not be economically and feasibly served by public sanitary sewerage facilities was placed in the regional land-use plan on soils which have only very slight, slight, or moderate limitations for development utilizing on-site soil absorption sewage disposal facilities. Within the areas shown for residential development by 1990, there are numerous small pockets of soils unsuited for development even with public sanitary sewers. These small areas can be avoided in most cases through proper subdivision design and placed in minor drainage-ways and local parks and open spaces.

Of the more than 1,085,000 acres of land used for agriculture in 1963, over 40%, or about 444,000 acres, was classified as prime agricultural land. The delineation of prime agricultural land, as noted above, was based on the regional soil survey. Urban expansion by 1990 within the region will require the conversion of more than 102,000 acres of agricultural land to urban use. The adopted regional land-use plan places all remaining agricultural lands into a recommended exclusive agricultural zone. In accordance with the regional development objectives and standards set forth above, nearly 423,000 acres, or about 95%, of the prime agricultural lands have been recommended for retention in agricultural use at least through 1990.

As noted above, the regional soils data provided an important input to the delineation of the primary environmental corridors. These high-value natural-resource corridors, which comprise about 17% of the total area of the region but which encompass almost all of the best remaining elements of the natural-resource base, were incorporated into the adopted regional land-use plan as a major plan element. The plan recommends that these corridors be refined as urban development continues in the region and that they be preserved and protected from encroachment by incompatible types of urban development. These corridors will also serve to provide the communities within the region with additional park and outdoor recreation areas.

Comprehensive watershed plans

The regional soil-survey data and interpretive analyses have also been extensively utilized in the Commission's series of comprehensive watershed studies. To date the Commission has prepared or currently has under preparation, comprehensive watershed-planning programs for four watersheds in the region: the Root River, Fox River, Milwaukee River, and Menomonee River watersheds. Comprehensive watershed studies are designed to produce for each watershed a long-range plan for the development of water-related community facilities, including integrated proposals for pollution abatement, drainage and flood control, land and water use, and park and public open-space reservation. As such, they are fully integrated into the ongoing regional planning program for land use, transportation facilities, and other public facilities and utilities.

An important part of each Commission watershed study is the development of a mathematical model, used to simulate the hydrologic and hydraulic performance of the river system under study. Each such simulation model is constructed from available information on the climate, topography, soils, land use, and hydraulic characteristics of the watershed. These factors are combined in the model through established hydrologic and hydraulic relationships. The model, one formulated, is calibrated to the specific watershed by using data on actual river performance, including high-water marks and stream gaging records. As the model is thus refined, a basic understanding of the specific hydrologic relationships of the watershed is obtained. The model then becomes a tool for forecasting river system performance, given, for example, a proposed or forecast change in one of the hydrologic input factors, such as land use. In the Commission's watershed studies, the hydrologic simulation model is used to simulate flood flows corresponding to selected recurrence intervals of 10, 50, and 100 years for conditions of present and planned future land use in the particular watershed under consideration. In this way floodlands can be delineated for use in conjunction with such public land regulatory devices as zoning and subdivision control.

Soils data are an important input to the development of a hydrologic simulation model. In the Commission's watershed studies, the watersheds are divided into hydrologic sub-basins. Detailed soils maps are then used to determine the predominant hydrologic soil group in each sub-basin. For this purpose all soil types occurring in the region have been classified into one of four hydrologic soil groups, A through D. The various hydrologic soil groups indicate the infiltration characteristics of the sub-basin soils, the group A soils having the highest infiltration rate and group D the lowest.

The hydrologic soil classification is used to determine the ratio of runoff to rainfall and thus assists in building the hydrologic model. As noted, the existing and proposed land uses also affect the amount of runoff. In view of the availability of the detailed soils data, the U.S. Soil Conservation Service Runoff-Curve-Number System was selected in the watershed studies as the most suitable method for calculating runoff resulting from a rainfall of given depth and duration (U.S. Dept. of Agriculture, Soil Conservation Service, 1972). This method assigns runoff-curve numbers to a range of hydrologic soil-cover complexes made up of combinations of hydrologic soil groups and agricultural land uses, and weighted average runoff curve numbers can be assigned to those sub-basins having mixed land use.

The detailed soils data have at least two additional applications in comprehensive watershed planning as conducted by the Commission. The soils data, in terms of its interpretations for flood hazard, are used in conjunction with the mathematical hydrologic simulation model to delineate accurately the 10-year recurrence interval flood inundation line along a stream system. Experience has shown that a strong correlation exists between such soil interpretations and the predictive 10-year recurrence interval flood. The soils can thus be used in a supplemental way in floodland delineation (S.W.R.P.C., 1969a). In addition, the detailed soils data are often used to assist in estimating the costs of proposed utility services. For example, in the Commission's Fox River watershed study, the soil maps and interpretive analyses were consulted in preparing cost estimates for the installation of several recommended public sanitary sewer systems. Where the proposed installations traversed soils having severe and very severe limitations for urban development utilizing sanitary sewers, higher unit cost factors were applied in preparing the estimate.

Regional sanitary sewerage system plan

The regional soil survey data and interpretive analyses have also been utilized in the regional sanitary sewerage-system planning program undertaken by the Commission. A major work element in this planning program is a technical analysis of the soils data with particular respect to that soils information having relevance for sanitary sewerage system planning. In particular, the areas proposed in the regional land-use plan to be developed for urban use and covered by soils suitable for septic-tank sewage-disposal system application and areas proposed to be developed for urban use and covered by soils unsuitable for this purpose have been mapped, measured, and tabulated by county, civil division, and subwatershed area. In addition, areas of bedrock outcrop, shallow bedrock, and high groundwater table have been mapped and analyzed as these factors may relate to the planning, design, and provision of sanitary sewerage facilities. The soil-survey data are also useful in evaluating sewage-treatment lagoon location and in the design of such lagoons, but are also most important in evaluating the feasibility of land disposal of treated sewage effluent through spray or furrow irrigation. All of these data serve not only as an aid in the system design but also in the preparation of cost estimates of various plan elements.

Thus, the detailed soils data continue to be invaluable to ongoing regional planning efforts. Proposed future regional planning programs, including programs designed to prepare a regional airport plan, a regional water-supply system plan, and a regional park and outdoor recreation plan, will also have to utilize extensively the detailed soils data.

Soils data and regional plan implementation

Each Commission planning report that recommends for adoption a regional or sub-regional plan element contains specific plan implementation recommendations to those federal, state, areawide, and local units of government that have the legal powers and financial means to implement most effectively the particular plan element under consideration. Certain of these plan implementation recommendations relate directly to, and often incorporate, the regional soil survey and its accompanying interpretive analyses (S.W.R.P.C., 1969b). Important among such implementation recommendations are those relating to the incorporation of soils data in local zoning regulations, land subdivision regulations, and health and sanitary regulations.

The soil survey and accompanying interpretations have, to date, been directly incorporated into local zoning ordinances in accordance with regional plan implementation recommendations in the following ways.

(1) Through the creation of special zoning districts related to certain kinds of soils, with particular emphasis in this respect on exclusive agricultural, conservancy, and certain types of residential zoning districts.

(2) Through the incorporation of special use regulations relating to certain kinds of soils.

(3) Through the delineation of zoning district boundaries and in the determination of special hazard areas, such as floodland areas.

Soils data have similarly been incorporated into local subdivision control ordinances, both in the form of general suitability clauses and in the form of special provisions relating to the control of grading operations, the design of drainage facilities, the removal of natural ground cover, and control of erosion and sedimentation.

Perhaps most importantly with respect to implementation of the adopted land-use plan has been the incorporation of soils data into local health and sanitary regulations. Such regulations serve in effect to prohibit the installation of septic-tank sewage-disposal systems on soils having high water tables, low permeability, or excessively high permeability, and on excessively steep slopes.

To date, the detailed soil-survey data and accompanying interpretations have been incorporated by specific reference into eight local zoning ordinances covering 1,050 square miles of area, or approximately 40% of the total land area of the region; into four local land-subdivision control ordinances covering 875 square miles of area, or about 33% of the total area of the region; into five local sanitary ordinances covering 1,990 square miles, or approximately 75% of the total land area of the region. Together these local land-use control ordinances, being based upon the same soils data as was the regional land use plan, contribute significantly to the implementation of the regional land use plan.

In addition, many miscellaneous uses of the soils data have contributed to regional land-use plan implementation, including the use of such data in land appraisal and assessment (Klingelhoets and Westin, 1954); in the development of cost estimates for rural and urban development proposals, street and highway location and design, storm-water drainage design (S.W.R.P.C., 1965) and specific site selection; and in the preparation of neighborhood unit development plans.

SUMMARY AND CONCLUSION

The regional and watershed planning programs in southeastern Wisconsin represent a unique effort to relate the preparation of areawide urban development plans to the natural resource base so that future development problems and accompanying deterioration of the regional environment may be avoided. Need exists in such a comprehensive regional planning program to examine not only how land and soils are presently utilized but also how this resource can best be used and managed. This requires an areawide soil suitability study which shows the geographic locations of the various kind of soils; identifies their physical, chemical, and biological properties; and interprets these properties for land-use and public-facilities planning. The resulting comprehensive knowledge of the character and suitability of the soils is one of the most important tools through which an adjustment of areawide urban development to the supporting resource base can be accomplished and is extremely valuable in every phase of the planning process.

The soils information converted to graphic form has application in any planning operation involving the spatial location of uses on the land. An example of such an operation is the preparation of a regional land-use plan by conventional planning techniques. In this application, areas covered by soils poorly suited to each of the various land uses are delineated on planning base maps so that these areas may be avoided in the spatial distribution of proposed land uses during plan synthesis. In this way, the detailed operational soils survey provides a new input into traditional plan synthesis operations and makes these traditional techniques far more powerful and effective. Other planning operations in which the soils data has been applied in graphic form include the preparation of zoning district maps, the design of detailed neighborhood unit development plans, the design of subdivision layouts, and the preparation of site plans. The graphic data has also been used in subdivision plat review, tax assessment, and development financing. Other uses readily suggest themselves in resource development and management, as well as in other functional areas of planning.

The soils data converted to numeric form have application in any planning operation involving mathematical computation and analysis. An example of such an operation is the determination of weighted coefficients of runoff. Such determination permits the soils data to be directly applied in drainage and storm-sewer design and thereby permits a significant improvement in existing design techniques. Another example of the application of the soils data in numeric form is in the determination of developable land area acreage figures. Such application greatly expedites the quantification of land-use planning data and the achievement of a balanced allocation of land to the various uses which meets the land-use demand growing out of the social, economic, and physical needs of the region and is at the same time properly related to the natural-resource base. The most important application of the numeric data, however, is in the newer land-use planning techniques, which utilize mathematical models in the synthesis of land-use and community facilities plans. Both the land-use design model, developed by the Commission to synthesize land-use plans, and the land-use simulation model, developed by the Commission to test land-use plans, require detailed operational soils data in numeric form for their operation. Indeed, such soils data provide the essential link between development costs and geographic location inherent in the construct of the models and essential to their application.

Soils information can comprise important inputs into the formulation of goals and objectives, the preparation of planning standards, the analysis of existing land use, plan synthesis, test, and evaluation, and, perhaps most important of all, plan implementation. In southeastern Wisconsin the soil information is comprising important inputs into the regional land-use and transportation planning program, into watershed planning programs, and into the community planning assistance programs presently underway.

Moreover, detailed operational soil-survey data can, if properly applied, provide the basis for many important day-to-day community development decisions by federal, state, and local units of government and private investors. Definitive soils data are essential to intelligent zoning, subdivision control, and official mapping at the local level of government, just as such data are essential to the preparation of a regional land-use plan, a regional transportation plan, a comprehensive watershed plan, or an intelligent farm-conservation plan. Since the soil surveys represent a basic scientific inventory, they provide valuable information needed for the planning, location, and design of highways, parks, land subdivisions, and sewage-disposal facilities, as well as for agricultural and forest land-use planning and management.

If soil properties, as revealed by a detailed operational soil survey, are ignored during either general or detailed plan formulation, not only will expensive obstacles to plan implementation occur, but irreparable damage may be done to the land and water resources of the community. A detailed operational soil survey, therefore, is one of the soundest investments of public funds that can be made.

REFERENCES

Klingelhoets, A.J. and Westin, F.C., 1954. Soil survey and land evaluation for tax purposes. *S. Dakota Agric. Exp. Sta. Circ.*, 109: 11 pp.

Klingelhoets, A.J., Carroll, P.H., Fox, R.E., Lee, G.B., Hole, F.D. and Milfred, C.J., 1968. Classification of Wisconsin soils. *Univ. Wisc. Coll. Agric. Life Sci., Spec. Bull.*, 12: 41 pp.

Shaler, N.S., 1891. The origin and nature of soils. *U.S. Geol. Surv. 12th Ann. Rep.*, I: 219–345.

Simonson, R.W., 1962. Soil classification in the United States. *Science*, 137: 1027–1034.

Soil Survey Staff, 1951. *Soil Survey Manual*. U.S. Dept. Agric. Handbook No.18, Govt. Printing Office, Washington, D.C., 503 pp.

Soil Survey Staff, 1973. *Soil Taxonomy: A Basic System of Soil Classification for Making and Interpreting Soil Surveys*. U.S. Dept. Agric. Handbook No.436, Govt. Printing Office, Washington, D.C.
Southeastern Wisconsin Regional Planning Commission, 1965. *Determination of Runoff for Urban Storm Water Drainage Design*. Technical Record, Vol.2, No.4, 19 pp.
Southeastern Wisconsin Regional Planning Commission, 1966a. *A Mathematical Approach to Urban Design*. Technical Report, No.3, 58 pp.
Southeastern Wisconsin Regional Planning Commission, 1966b. *Soils of Southeastern Wisconsin*. Planning Report, No.8, 403 pp.
Southeastern Wisconsin Regional Planning Commission, 1969a. *Floodland and Shoreland Development Guide*. Planning Guide, No.5, 199 pp.
Southeastern Wisconsin Regional Planning Commission, 1969b. *Soils Development Guide*. Planning Guide, No.6, 247 pp.
U.S. Department of Agriculture, Soil Conservation Service, 1972. *Hydrology*. S.C.S. National Engineering Handbook, Section 4, 586 pp.
U.S. Department of Commerce, Bureau of the Census, 1971. *General Demographic Trends for Metropolitan Areas, 1960 to 1970*. PHC(2)-1, 120 pp.

Geoderma, 10 (1973) 27–34

TWO DECADES OF URBAN SOIL INTERPRETATIONS IN FAIRFAX COUNTY, VIRGINIA

D.E. PETTRY and C.S. COLEMAN

Department of Agronomy, Virginia Polytechnic Institute and State University, Blacksburg, Va. (U.S.A.)
Fairfax County soil scientist, Fairfax, Va. (U.S.A.)

(Accepted for publication July 24, 1973)

ABSTRACT

Pettry, D.E. and Coleman, C.S., 1973. Two decades of urban soil interpretations in Fairfax County, Virginia. *Geoderma*, 10: 27–34.

Two decades of soil interpretation in rapidly expanding Fairfax County, Virginia, caught in the urban sprawl of metropolitan Washington, D.C., have dramatically demonstrated the essential role of soil information as a basic tool for diversified land utilization. Merging the traditionally separated roles of soil interpretation for agricultural and urban sectors, the Fairfax program has pioneered the role of the urban soil scientist and paved the way for widespread usage of soil surveys in urban areas. Emerging from isolated problem solving, the program has integrated with existing agencies and other disciplines on a functional basis to provide protection and economic savings to the general public.

INTRODUCTION

Almost two decades have passed since Fairfax County, Virginia, became the first area in the United States to hire a soil scientist for multi-purpose soil interpretations on a county basis (Smith, 1960). Although rather commonplace today, the inauguration of an urban soil interpretation program two decades ago to utilize a soil survey for other than agronomic purposes was a bold venture into the unknown. Doomed by some to a short life, while predicted by others as a forerunner of future developments in the utilization of soil surveys, the Fairfax County urban soil program has flourished since 1955. This report examines the developments of this unique program in opening new horizons in soil surveys.

Caught in the urban sprawl of metropolitan Washington, D.C., Fairfax County, Virginia, has been transformed from a rural county with a population of 40,000 in 1940 to a sprawling urban complex of some 422,000 persons in 1972. The county comprises some 414 square miles in the northeastern part of Virginia and it is situated in the northern Piedmont and Coastal Plain physiographic provinces (Fenneman, 1938).

The initiation of the soil survey by the Virginia Agricultural Experiment Station in cooperation with the Soil Conservation Service was unique in 1953 since many of the requests precipitating the survey came from non-agricultural agencies (Robinson et al., 1955).

Frustrated planners attemped to utilize a 1915 vintage soil survey, with a scale of one inch equals one mile, for land use planning to accommodate the population influx. These planners were instrumental in obtaining a modern soil survey. The survey commenced in July, 1953, using current aerial photographs, with a scale of four inches equals one mile (1:15,840), as a base map. The field work, which was completed in 1955, established the framework for building the urban interpretation program. Approximately 100 square miles of the county were not included in the survey because the county deemed it to be essentially urbanized in 1953 with public water and sewer facilities already in this area. A preliminary soil report containing multi-purpose interpretations for each soil unit was published by the Virginia Agricultural Extension Service in 1956 and was revised in 1958 (Porter, 1956, 1958). A report on the chemical and physical properties of dominant soils in the county was published by the Virginia Agricultural Experiment Station in 1960 (Obenshain et al., 1960). The Soil Survey Report containing maps was published by the Soil Conservation Service in 1963 (Porter et al., 1963).

DISCUSSION

Integration of the soil-survey information into the existing county agencies as a basic part of their functional operations occurred over a short transition as benefits of this information became increasingly apparent. Planners readily incorporated soil maps and multi-purpose interpretations into their planning and zoning programs. Soil maps soon became a basic format for the development of comprehensive land-use plans. The maps were combined with other definitive criteria for use in evaluating property for tax assessment. Soil maps were reproduced to enlarged scales compatible with county property maps, and methods were developed to effectively handle and store soil data for immediate recall, including use of computers. Increasingly, questions were raised in the county when obvious mistakes occurred in site selections for schools, public buildings, and other facilities where costly errors due to deleterious soil conditions could have been avoided by using the soil survey data. Widespread publicity of these avoidable costly ventures tended to focus attention on the urban soil program and also inform the general public of the program and what it had to offer. On the other hand, the publicity served to put the urban soil program on trial in the public "limelight" where each major soil evaluation received the scrutiny of a wide populace determined to test this new operation. The economic magnitude of decisions, made largely on the basis of soil evaluations, often meant that vast sums of money and unmistakable pressures were thrust upon the urban soil scientist. Accountability for soil evaluations and the subsequent consequences of such decisions usually occurred shortly, often in a matter of hours following an excavation.

Armed with a detailed soil survey and laboratory data on the major mapping units, which were reinforced with a knowledge of each unit that became more definitive with each interpretation, the urban soil scientist soon proved that a soil survey had a definite place in an urban county. The soil scientist's position was made a direct responsibility of the county administrator to avoid possible bureaucratic entanglement within the administrative framework which could hamper operations.

Environmental health

When the soil interpretation program commenced in Fairfax County, one of the major soil problems was failing septic systems (Fig.1). Prior to the soil survey, the health department had depended almost entirely on percolation tests to determine the feasibility of the site for septic tank installation. Frequently, soils would pass the percolation test in dry weather, but the system would fail in wet weather. Percolation data and knowledge of areas that failed could not be extended to other areas of the county since the information was not identified with specific soils. Septic system drainfields were often installed at arbitrary depths without recognizing the differences between soil horizons. The detailed soil survey and interpretations provided a valuable tool for sanitarians to use in evaluating sites for septic systems and in designing the kind and size of individual disposal systems for each site. The Fairfax County Sanitation Division developed a method of using the detailed soil survey as a basic guide in their work (Clayton et al., 1959). Basically, the program consisted of an initial general evaluation of the land and a more detailed examination of apparently suitable areas. The soil survey indicated soil areas with seasonally high water tables, sites subject to periodic flooding and soils that were too shallow or steep to assimilate sewage. Percolation rates were conducted on soils rated as satisfactory or better in order to determine the proper drainfield size for each specific site.

Fig.1. Darker colored grass and effluent outline the septic tank drainlines of a failing system located on a poorly suited soil.

The basic soil-evaluation system is still being employed, with marginal soil areas getting additional on-site investigations by the urban soil scientist to provide very detailed parameters to the sanitarians. Drastic reduction in septic-tank system failures resulted from this program. A recent study by the Fairfax Health Department (R. Miller, personal communication, 1972) of subdivisions with individual septic systems installed over the past 15 years revealed an average failure rate of less than 5%. Follow-up investigations of the very low percentage of failing systems indicated many failures were due to overloading the system beyond original design criteria due to additions of dishwashers, garbage disposals, and other water-contributing sources. Based on this study, it is predicted that utilizing soil information, septic systems installed in 1956 may have an estimated effective life of 20 to 30 years or more, rather than the initial estimate of 15 to 20 years.

Interfacing of soil information with environmental health programs occurred at an increasing rate as the urban soil scientist was requested to locate suitable soil areas for sewage lagoons and sanitary landfills. Indiscriminate dumping of solid waste in the past (Fig.2) and recent anti-pollution ordinances provided the impetus to dispose properly of the increasing amounts of solid waste in centralized areas without polluting the environment. The soil scientist has served a key rôle in the selection of suitable soil areas that have sufficient areal extent, depth, and proper physical, chemical, and mineralogical parameters to adequately assimilate the solid waste with minimal contamination hazards. The increased

Fig.2. Numerous indiscriminate dumps of solid waste littered many rural roads and vividly demonstrated the need for proper sanitary landfills.

data base developed for each soil mapping unit over the years has provided ready access to information used to screen areas prior to on-site investigations of apparently suitable areas.

Flood plain development

In the rapid-building era of the 1950's many areas previously in fields and forests succumbed to asphalt and concrete roads and pavements, and they were replaced by shopping centers and subdivisions. Some of the structures were constructed on flood plains and on soil areas subject to occasional ponding. Initially, when there were only one or two subdivisions in a watershed, occasional minor flooding and wet basements resulted. However, as building increased in watersheds, the surface runoff increased and seasonal flooding became an increasing threat (Fig.3).

Fig.3. Flood waters delineate the flood plain of a small stream in a rural area clearly illustrating how soils often differ markedly in short distances.

Severe flood damage in 1956 precipitated costly detailed engineering and hydrological studies on the major flood-plain areas of the county. Resulting delineations of the 100-year flood plains of major streams and tributaries were in very close agreement with boundaries delineated by the soil scientist on the basis of soil characteristics. The county enacted an ordinance prohibiting building developments within flood-plain boundaries. This ordinance stated that on streams which have not had an engineering or a U.S. Geological survey completed, the limits of the flood plain will be based on the soil maps as interpreted by the county soil scientist. Once the flood-plain areas were determined, many were purchased by the county and maintained as natural "green-belt" areas for parks and recreation.

Building development

In the winter of 1961, after several unusually deep snows had melted, landslides occurred in several parts of the county. Soil investigations of these areas showed that soil conditions in each case were similar. Slides were occurring in plastic, sticky, gray marine-clay sediments that were overlain by porous, sandy soils. Developments which removed vegetation, altered runoff characteristics, and excavated bases of supporting slopes accelerated the slides. Unfortunately, several houses (Fig.4), schools, and other structures were affected and some were destroyed.

Fig.4. A new house exhibiting severe structural failure due to differential soil movement of the unstable soil.

Areas in the county subject to slides when disturbed were delineated by the soil scientist. The present policy of Fairfax County prohibits building permits in these areas until the builder provides a method of correcting the slide susceptibility that satisfies the Public Works Department. The experience vividly demonstrated the need to understand the nature of deeper subsurface materials below what was traditionally considered to constitute soil. Fortunately, these soil areas had been separated by means of phases in the soil survey and they could thus be readily marked throughout the county.

The county school board policy requires soil investigations of all tentative school sites. Usage of soil information for site selection and design criteria have become a routine procedure for the county educational system which in 1971 had become the 16th largest school system in the United States (Fairfax Annual Report, 1971) and contained 170 schools by 1973. Due to the very rapid growth and anticipating new school construction, the soil scientist was requested to help select and evaluate school sites which were purchased by the county several years in advance of actual construction. Not only were costly construction practices due to adverse soil conditions avoided, but large economic savings were realized due to increasing land values.

Fig.5. Cracked pavement and "pot-holes" serve as a constant reminder of the seemingly endless, costly maintenance problems resulting from road construction on poorly suited soils.

Recognition of soils with expansive clayey horizons affecting the construction of roads and structures and location of these areas on the soil maps have avoided costly mistakes and enabled proper design. Construction specifications for these soil areas have reduced costly maintainance of roads (Fig.5) and have diminished cracked foundations.

CONCLUSIONS

The Fairfax County soil program has demonstrated over the past two decades that, although traditionally separated, many of the same soil factors that affect agriculture also influence urban development. It is somewhat ironic that much of the soil interpretation has been conducted in the 100 square mile area that was not mapped in the survey due to the urban development at that time. Rapid urbanization and alteration of the surface topography complicated soil interpretations. Essentially, the area had to be mapped in order to make soil interpretations. Although "hindsight" is quite accurate, foresight in anticipating future developments and diverse uses of soil information is quite limited.

The various applications of soil information in a developing urban complex have demonstrated the need of detailed, accurate soil delineations and accurate recording of drainage systems and topographical features on the soil maps. Experience has shown that rapid urbanization can often disguise or alter these features which have been observed to greatly influence the subsequent environment and land use. The importance of recognizing and understanding the nature of deeper substrata materials and their influences on the surface soils and land usage cannot be overemphasized. Susceptibility of sliding, depth to bedrock and physical nature of the material are very important in intensive land utilization. The relationship of soil bodies to other environmental parameters and the dynamic relationship of the ecosystems have been demonstrated to be very significant.

REFERENCES

Clayton, J.W., Kennedy, H., Porter, H.C. and Devereux, R.E., 1959. *Use of Soil Survey in Designing Septic Disposal Systems. Res. Bull. 509.* Virginia Agricultural Experiment Station, Blacksburg, Virginia, 14 pp.

Fairfax County Annual Report, 1971. Division of Public Affairs of Fairfax County, Virginia, 56 pp.

Fenneman, N.M., 1938. *Physiography of Eastern United States.* McGraw-Hill, New York, N.Y., 714 pp.

Obenshain, S.S. and Porter, H.C., 1960. *Chemical and Physical Properties of Fairfax County Soils. Res. Rep. 41.* Virginia Polytechnic Institute, Blacksburg, Virginia, 34 pp.

Porter, H.C., 1956. *Soils of Fairfax County. Extension Rep. Ser. 3.* (1st Ed.) Virginia Polytechnic Institute, Blacksburg, Virginia, 139 pp.

Porter, H.C., 1958. *Soils of Fairfax County. Extension Rep. Ser. 3.* (2nd ed.). Virginia Polytechnic Institute, Blacksburg, Virginia, 167 pp.

Porter, H.C., Derting, J.F., Elder, J.H., Henry, E.F. and Pendleton, R.F., 1963. *Soil Survey of Fairfax County, Virginia.* U.S. Department of Agriculture, Soil Conservation Service, U.S. Government Printing Office, Washington, D.C., 103 pp.

Robinson, G.H., Porter, H.C. and Obenshain, S.S., 1955. The use of soil survey information in an area of rapid urban development. *Soil Sci. Soc. Am. Proc.* , 19:502–504.

Smith, V.W., 1960. A relator's views about soil surveys. *Soil Conserv.*, 26:106–108.

Geoderma, 10 (1973) 35–45

SOIL SURVEY FOR URBAN DEVELOPMENT*

J.D. LINDSAY, M.D. SCHEELAR and A.G. TWARDY

Soils Division, Research Council of Alberta, Edmonton, Alta. (Canada)

(Accepted for publication July 11, 1973)

ABSTRACT

Lindsay, J.D., Scheelar, M.D. and Twardy, A.G., 1973. Soil survey for urban development. *Geoderma*, 10: 35–45.

As an aid to urban planning, a detailed soil survey is underway in the new Mill Woods district, City of Edmonton. Five soil associations – Ellerslie, Mill Woods, Argyll, Hercules, and Beaumont – are being mapped in the area. These soils represent the Chernozemic, Solonetzic, and Gleysolic Orders in the Canadian System of Soil Classification. Ellerslie and Beaumont soils are a good topsoil source and will support plant growth well. Argyll soils have poor surface drainage and being saline may present problems in lawn establishment and trafficability. The Mill Woods soils are intermediate between Ellerslie, Beaumont and Argyll associations. Hercules soils are poorly drained and may be saline; their suitability for landscaping purposes is limited.

From an engineering or construction standpoint, the soils in the Mill Woods area may present some problems related to shrink–swell potential and concrete corrosion. The Argyll, Mill Woods and Hercules soil associations are characterized by a relatively high sulfate content which represents a potential concrete corrosion hazard in the area.

The soil map also indicates areas of poorly drained soils which will require special engineering practices before development can be initiated.

INTRODUCTION

Historically, requests for soil information originated with individuals whose interests were related to the production of food, fibre or timber products. More recently, however, an increased awareness of the suitability or unsuitability of soils for other uses has become apparent. The unprecedented growth of urban communities expanding to new areas has resulted in requests for soil information concerning this particular alternate land use.

Recently, 6,000 acres of land were annexed adjacent to the City of Edmonton, Alberta, for urban expansion. According to the report of the City Planning Department (1971), this new community called Mill Woods will eventually accommodate 120,000 people and have a development time span of more than 20 years.

Development of the area began in 1970 and to assist with the planning, a high intensity (detailed) soil survey was carried out on 260 acres of land in that year. Subsequently, in 1971 and 1972, 500 and 320 acres, respectively, were mapped by soil survey.

City planners indicate the areas of developmental priority within the Mill Woods district and the soil-survey program is scheduled to provide information on the areas of immediate

* Research Council of Alberta, Contrib. No. 640.

concern. Each year a soil map and soil interpretation maps of topsoil suitability, potential concrete corrosion hazard and soil drainage are prepared.

This paper describes the methodology used in conducting the soil survey and presents some of the interpretations based on the morphological, chemical and physical properties of the soils.

METHODS

The area to be mapped is carefully surveyed into a grid system in which stakes are established along lines spaced at 250-ft. intervals. This step ensures that adequate control is provided for soil mapping and soil sampling procedures. The aerial photographs used in the study are at a scale of 500 ft. to 1 inch (1:6,000).

A truck-mounted coring drill is used for the inspection of the soils. Cores, to a depth of 7 ft., are obtained at each of the staked sites in the grid system.

The soils are classified at each survey stake, using the cores, but sampled at only every second stake. Thus soil samples are obtained at 500-ft. intervals throughout the area.

In the laboratory, emphasis is given to the determination of water-soluble salts and electrical conductivity on a saturation extract, soil reaction (pH), particle size distribution, Atterberg limits, total nitrogen and total carbon.

RESULTS

Soils and surficial deposits

The soils of the area were mapped by Bowser et al. (1962) at a scale of 1 inch to 2 miles (1:126,000). This soil mapping was carried out as part of a basic data inventory program. Because of map scale, however, the information has limitations when used in the context of detailed urban planning. Consequently, to provide the planners with as much information as possible the soil map for the present study is published at a scale of 1 inch to 250 ft. (1:3,000).

The portion of the Mill Woods district soil mapped to date is developed primarily on fine-textured lacustrine clay. This area is part of Glacial Lake Edmonton (Bayrock and Hughes, 1962) which covered much of the Edmonton district during Late Pleistocene time. Soils developed from glacial till are of minor occurrence in the area. According to drill records (Hardy and Associates, 1971) the stratigraphic sequence of material is variable, but portions of the area consist of 12 ft. of lacustrine material over 10 ft. of clay loam till overlying bentonitic shales and sandstones of the Edmonton Formation.

Generally the topography of the area is fairly smooth with slopes seldom exceeding 2%. Low-lying depressional areas are a characteristic feature of the landscape.

The soils are mapped on a soil association basis. Each soil association is a group of related soil series developed on a particular parent material. The soil association, therefore, represents a combination of natural features including the kind of landscape, the surface color of the soil and the dominant soil textures.

The map units represent a portion or a segment of a soil association and are composed of one or more soil series. Different map units are separated on the basis of different proportions of soil series occurring within the association. These are indicated as being dominant or significant in the association.

The use of soil associations appears suited to this study because the fine divisions included in the soil classification system can be recognized, although in some cases their intimate occurrence in the landscape precludes delineation on the soil map. This is not regarded, however, as a shortcoming of the mapping procedure since most of the inseparable units, although significant from a classification standpoint, are not sufficiently different with respect to soil properties to affect their use for urban development.

Soils are classified according to the System of Soil Classification for Canada (Canada Department of Agriculture, 1970).

Significant and dominant subgroups of the various soil associations employed in the Mill Woods district are shown on the soil map legend (Fig.1). This map represents only a portion of the area covered by soil survey in the Mill Woods district.

The Ellerslie association consists of moderately well to imperfectly drained Chernozemic soils developed from lacustrine sediments. These soils are characterized by a thick dark colored, granular and friable Ah horizon. The B horizon is friable whereas the C is moderately calcareous and weakly saline. Soils of this association range in texture from silty clay to clay.

Soils of the Mill Woods association are moderately well to imperfectly drained Solonetzic soils developed from fine textured lacustrine material. They are usually found in areas of groundwater discharge where salts have been brought near the surface by a fluctuating water table. The Ah horizon of the Mill Woods soil is thinner than that of the Ellerslie association. They have a hard B horizon which becomes a sticky, gelatinous mass when wet. The soils are characterized by a high salt content in the lower portion of the profile.

The Argyll association is comprised of imperfectly drained alkaline Solonetz soils that are extremely dense and intractable. These impermeable soils have a thin Ah horizon overlying a dense B horizon. Significant amounts of soluble salts occur in the B and C horizons. The hard compact B horizon limits the penetration of water, air and roots. Argyll soils are developed from silty clay to heavy clay lacustrine deposits.

The Hercules association includes those soils that are saturated or are under reducing conditions continuously or during some period of the year. They are poorly and very poorly drained soils found in areas of groundwater discharge. An accumulation of relatively undecomposed peat occurs at the surface of some of these soils. Soluble salts may occur in all horizons of the Hercules soils which are developed from lacustrine silty clay and clay.

Beaumont soils are moderately well drained Chernozemic soils developed from loam to clay loam till. These soils are characterized by a thick, dark colored, granular and friable Ah horizon. The B horizon is friable whereas the C is moderately calcareous and weakly saline.

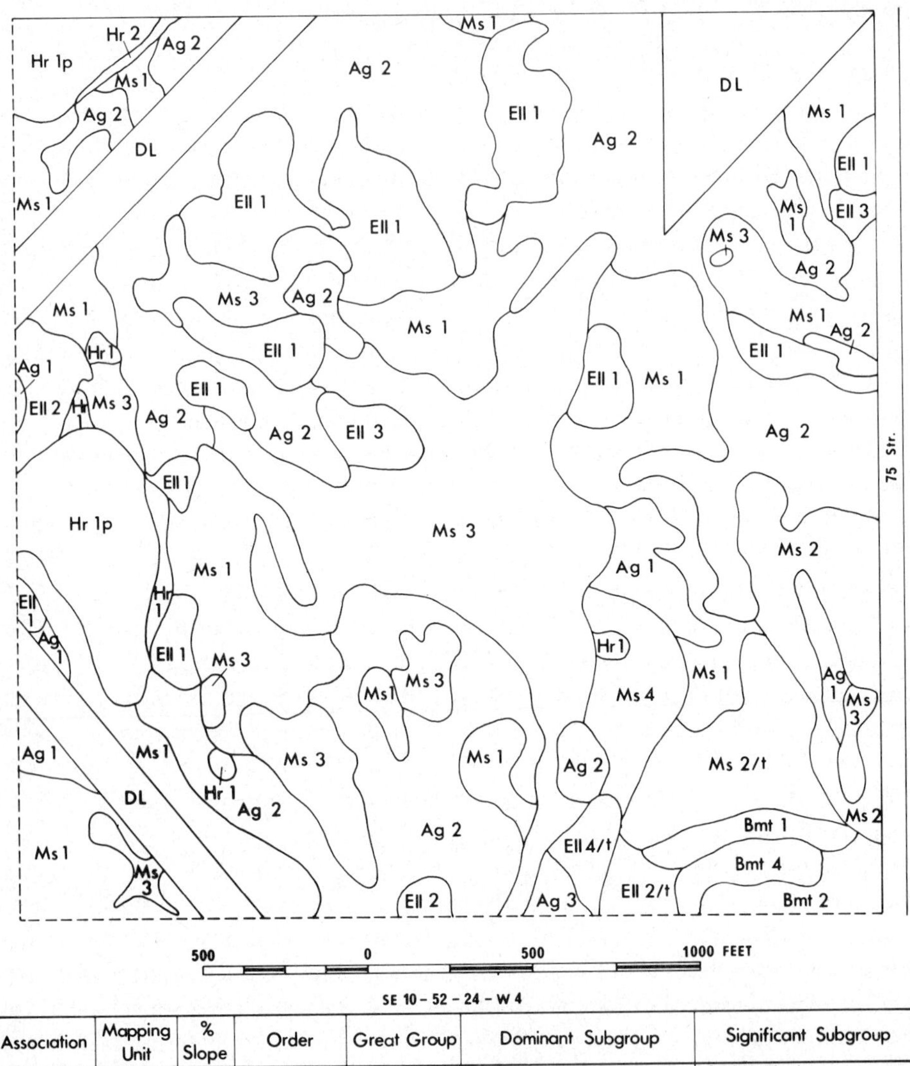

Association	Mapping Unit	% Slope	Order	Great Group	Dominant Subgroup	Significant Subgroup
Ellerslie	Ell 1	0 - 2	Chernozemic	Black	Eluviated Black	Orthic Black
	Ell 2	0 - 2			Solodic Black	Black Solod
	Ell 3	0 - 2			Gleyed Eluviated Black	
	Ell 4	2 - 5			Eluviated Black	Orthic Black
Mill Woods	Ms 1	0 - 2	Solonetzic	Solonetz and Solod	Black Solonetz	Black Solod
	Ms 2	0 - 2			Black Solod	Black Solonetz
	Ms 3	0 - 2			Gleyed Black Solonetz	Gleyed Black Solod
	Ms 4	2 - 5			Black Solod	
Argyll	Ag 1	0 - 2		Solonetz	Alkaline Solonetz	
	Ag 2	0 - 2			Alkaline Solonetz	Gleyed Alkaline Solonetz
	Ag 3	0 - 2			Gleyed Alkaline Solonetz	
Hercules	Hr 1	0 - 2	Gleysolic	Humic Gleysol	Orthic Humic Gleysol	
	Hr 2	0 - 2			Saline Rego Humic Gleysol	
Beaumont	Bmt 1	0 - 2	Chernozemic	Black	Eluviated Black	Orthic Black
	Bmt 2	0 - 2			Solodic Black	Black Solod
	Bmt 4	2 - 5			Solodic Black	Black Solod

p - peaty phase   DL - Disturbed Land   /t - overlying till

Fig.1. Soil map of a portion of the Mill Woods district.

Soil interpretations

From the basic soil survey data it is possible to make predictions of performance for the soils, based on soil morphology and associated soil physical and chemical properties. In the Mill Woods area there are two main uses for which the soils of the area will be required – lawns and landscaping, and as a construction material.

The data in Table I provide some indication of the suitability of the surface soils for landscaping. These data represent mean values for soil samples collected at some 116 sites in the area.

In descending order of topsoil suitability, the soil associations in the Mill Woods area can be ranked as Beaumont, Ellerslie, Mill Woods, Argyll and Hercules. This ranking is based on an evaluation of the Ah horizon in terms of organic matter content, total nitrogen content, water-soluble salt content, surface or internal drainage, texture, and permeability.

The Ellerslie and Beaumont associations have few properties that limit plant growth. The Argyll association, however, has fairly severe limitations because of its thin Ah horizon, relatively low organic matter content, moderate salinity, and low permeability. Rainwater tends to remain on the surface of these soils for relatively long periods to be removed only by evaporation. Such a phenomenon tends to result in the upward movement of salt-bearing groundwater (Cairns and Bowser, 1969). The soils of the Mill Woods association have moderate limitations for plant growth. They do not have the extremely undesirable physical and chemical properties of the Argyll association but at the same time the data indicate they are inferior to Ellerslie soils.

Areas of Hercules soils are poorly to very poorly drained. Such soils present vegetative rooting problems due to wetness. Also, salts have been brought near the surface by groundwater discharge which serves to further limit the suitability of these soils for landscaping and lawn establishment.

The Solonetzic soils, particularly those of the Argyll association, may present problems with respect to compaction and trafficability. These impermeable soils will puddle and compact under excessive pedestrian traffic at high moisture contents. School yards and

TABLE I

Mean and standard deviation values for some chemical and physical properties of the surface horizon (topsoil) of the soil associations in the Mill Woods district

Soil association	Thickness (cm)	pH	Organic matter (%)	Nitrogen (%)	Electrical conductivity (mmhos/cm)	Sulfate (%)	Texture
Ellerslie	24±9	6.4±0.2	12.2	0.69	0.4±0.1	0.00	silt loam
Mill Woods	17±5	6.4±0.1	11.2	0.59	1.1±0.4	0.02±0.07	loam
Argyll	12±4	6.2±0.4	8.7	0.49	1.5±0.8	0.03±0.04	clay loam
Hercules	22±11	6.8±0.1	10.9	0.67	2.0±1.7	0.06±0.03	clay loam
Beaumont	29±11	6.6	12.5	0.56	–	–	clay loam

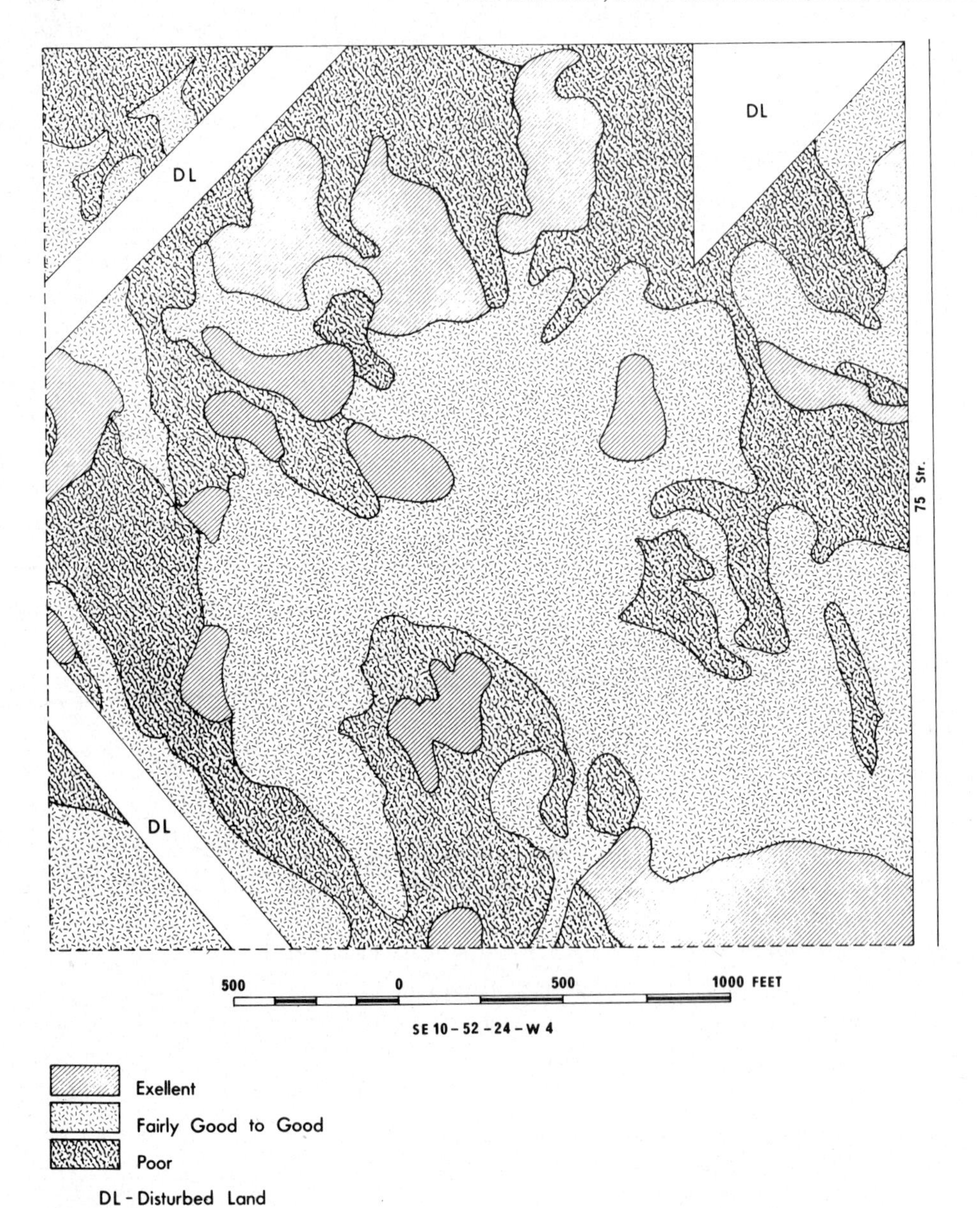

Fig. 2. Soil suitability for landscaping.

playgrounds will be particularly susceptible to this type of problem.

In the Mill Woods area considerable care must be taken in preparing the land for landscaping. Since subsoil salinity is characteristic of the Argyll, Mill Woods and Hercules soil associations, every precaution must be taken to ensure that the subsoil material is not left at the surface following construction. An interpretive soil map showing the suitability of the surface soils for landscaping purposes in a portion of the Mill Woods area is shown in Fig. 2.

TABLE II

Mean and standard deviation values for some chemical properties of the subsoil horizon (C) of the soil associations in the Mill Woods district

Soil association	pH	Electrical conductivity (mmhos/cm)	Sulfate (%)	m. equiv./litre		
				sodium	magnesium	calcium
Ellerslie	7.6±0.3	1.9±1.1	0.06±0.06	8.2±8.7	5.3±6.3	11.7±9.0
Mill Woods	7.6±0.2	4.3±2.0	0.22±0.14	33.9±29.4	15.8±11.0	20.4±7.5
Argyll	7.7±0.2	7.5±2.4	0.44±0.18	87.1±53.4	24.9±10.8	20.8±3.5
Hercules	7.6±0.3	4.7±2.9	0.24±0.20	41.4±38.7	18.5±19.0	17.0±7.3
Beaumont	7.7±0.3	1.7±1.2	0.04±0.05	8.9±14.3	3.9±4.1	10.3±10.0

From an engineering and construction standpoint, the soils in this area present a number of problems with regard to urban development. One major concern is the potential corrosion of concrete structures and underground conduits because of subsoil salinity. Some of the subsoil chemical properties of the soil associations are shown in Table II.

The mean sulfate content in the subsoil (C horizon) ranges from 0.04% in the Beaumont association to 0.44% in the Argyll association. Corresponding mean electrical conductivity measurements of a water extract from these soils are 1.7 and 7.5 mmhos/cm, respectively.

The principal soluble salt in the soils of the Mill Woods area is sodium sulfate, with magnesium sulfate and calcium sulfate also occurring to a significant extent. Pawluk and Bayrock (1969) and Swenson (1971) also report the dominance of these salts in some of the soils of the Canadian prairie region.

The *Concrete Manual* of the United States Bureau of Reclamation (1966) recognizes the following concrete corrosion categories:

Negligible attack: $<$0.10% sulfate in soil
Mild but positive attack: 0.10–0.20% sulfate in soil
Considerable attack: 0.20–0.50% sulfate in soil
Severe attack: $>$0.50% sulfate in soil

Using the above standards as a guideline, the potential corrosion hazard associated with the soil associations in the Mill Woods area ranges from negligible to mild in the Beaumont and Ellerslie associations, mild to considerable in the Mill Woods association, considerable to severe in the Argyll association, and considerable in the Hercules association.

Swenson (1971) has outlined in some detail the precautions that should be taken where concrete structures are to be placed in a sulfate soil environment. Such measures include the following: use of sulfate-resisting cement, a low water-cement ratio, high cement content, air-entrainment, waterproof coatings, drainage facilities, and special reinforcing cover.

An interpretive soil survey map showing the various areas of potential concrete corrosion in a portion of the Mill Woods area is shown in Fig.3.

Physical properties of the soil are of special interest to engineers because they affect design, construction, and maintenance of structures. In the Mill Woods area a number of sites were sampled specifically for engineering tests. The results of the analyses are shown in Table III.

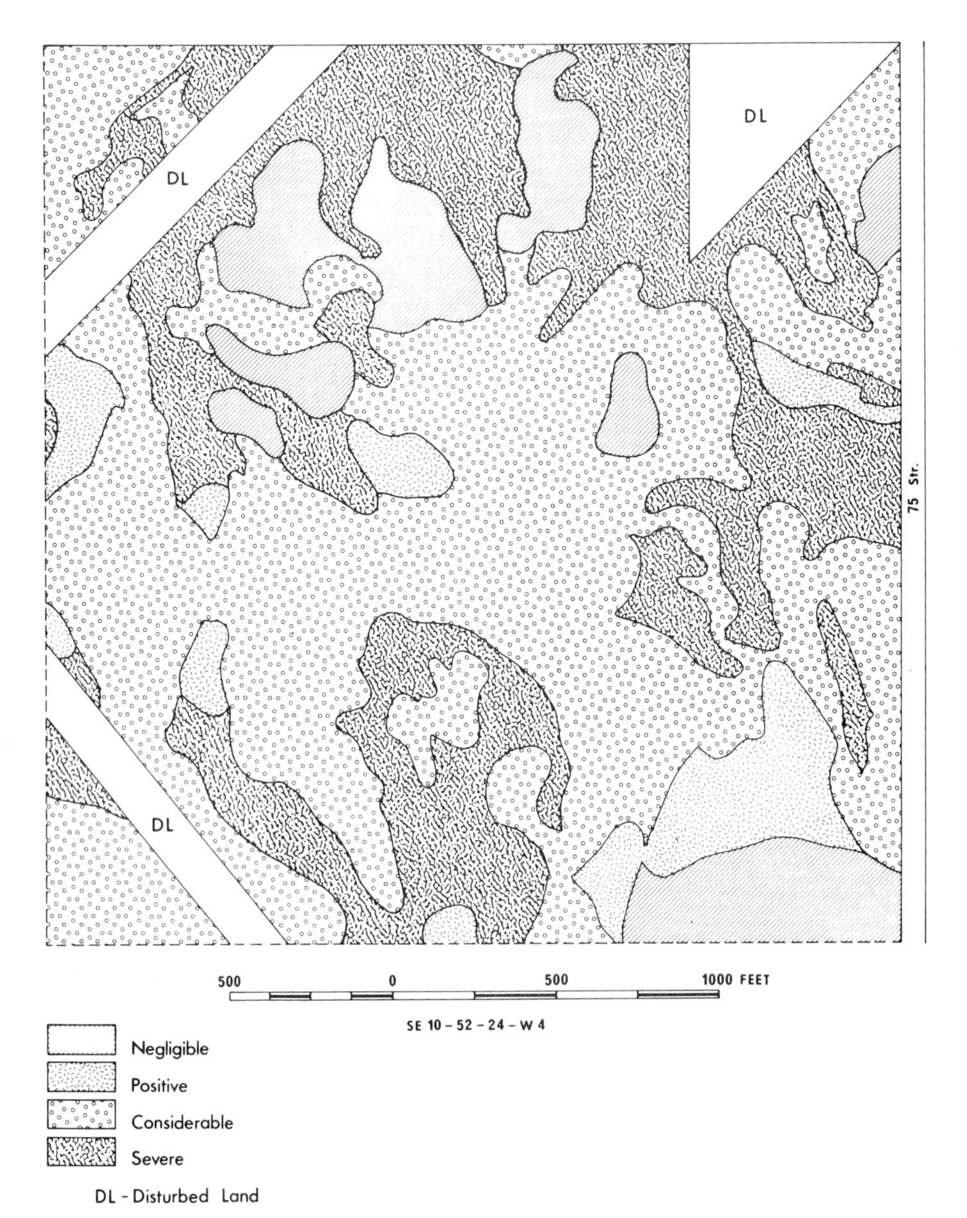

Fig.3. Potential concrete corrosion hazard of the soil associations.

Since a major portion of the Mill Woods area mapped to date is mantled to a depth of at least 7 ft. with lacustrine clay, the soil associations differ very little in so far as the engineering data are concerned. The exception is the Beaumont association which is developed from till rather than lacustrine material. The subsoils of the Ellerslie, Mill Woods, Argyll, and Hercules associations usually are classified as CH in the Unified System and A-7-6 in the AASHO system of classification, whereas the Beaumont association is CL and

TABLE III

Engineering test data for the subsoil (C horizon) of representative soil profiles of the soil associations from the Mill Woods district

Soil association	Depth from surface (cm)	Per cent passing sieve							Per cent smaller than				Liquid limit	Plasticity index	Activity No.	Textural classification		
		1 in.	3/4 in.	5/8 in.	No. 4	No. 10	No. 40	No. 200	.05 (mm)	.005 (mm)	.002 (mm)	.001 (mm)				AASHO	Unified	USDA
Ellerslie	50–100	100	100	100	100	100	96	70	69	48	38	34	43	21	0.6	A7-6	CL	SiC
	100–150	98	98	98	97	97	97	86	83	66	43	38	53	26	0.5	A7-6	CH	SiC
Mill Woods	50–100	100	100	100	100	100	99	94	93	79	61	51	66	37	0.6	A7-6	CH	SiC-HC
	150–225	100	100	100	100	100	100	96	95	78	52	41	60	33	0.6	A7-6	CH	SiC-HC
Argyll	90–140	100	100	100	100	100	98	88	87	74	65	62	64	39	0.6	A7-6	CH	SiC-HC
	165–225	100	100	100	100	100	96	74	72	54	40	33	61	34	0.9	A7-6	CH	SiC
Beaumont	63–180	100	100	100	100	99	94	72	70	42	33	28	34	12	0.4	A6	CL	CL
	180–250	100	100	100	100	99	93	68	67	40	32	29	33	13	0.4	A6	CL	CL

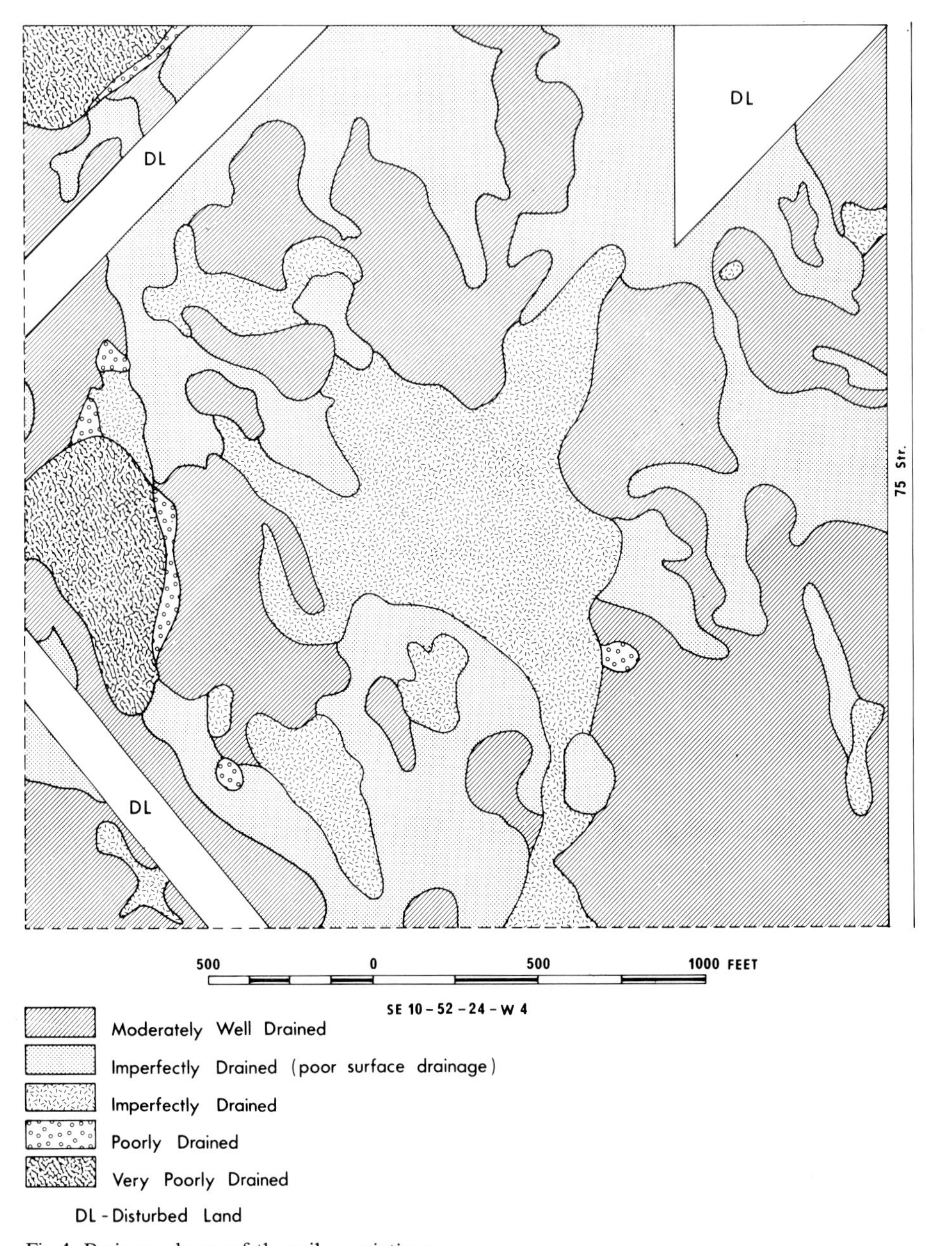

Fig.4. Drainage classes of the soil associations.

A-6. Such material, particularly the lacustrine sediments, is characterized by a high content of plastic clays. Arshad (1966), in a study of the soils of this area, indicates montmorillonite as the dominant clay mineral with illite and kaolinite next in abundance. Montmorillonite is an expansible clay and soils characterized by this clay mineral have a high shrink–swell potential and are subject to fairly large volume changes with change in moisture content.

The area of poorly drained soils, the Hercules soil association, is of similar texture to other soils of the area but wetter. These soil areas have water tables near the surface and may present problems in bearing strength and drainage for structures. A map of a portion of the Mill Woods district showing the various soil drainage classes is shown in Fig.4.

Special engineering practices are required in the poorly drained areas where artificial drainage, use of piling or addition of fill material may be required before construction can be initiated.

CONCLUSIONS

Requests for detailed or high intensity soil surveys are rapidly increasing. Such surveys can be used for planning a wide assortment of facilities ranging from homes and industrial plants to schools and playgrounds. The cost-benefit ratio of such surveys has been estimated at 1 to 100 (Klingebiel, 1966).

The information provided by the soil survey of the Mill Woods district of Edmonton has aided in the formulation of construction practices, particularly in regard to the specifications for the type of concrete to be used in the area. At the same time new home owners benefit from the soils information in that recommendations can be made with regard to the preparation of the soils for lawns and landscaping.

It should be mentioned, however, that soil survey information is not meant to eliminate the need for deep borings for specific structures. The erection of high-rise towers and large buildings will require on-site investigation; the soil survey, however, aids in determining where the deep borings should be made and where the buildings should be sited.

REFERENCES

Arshad, M.A. and Pawluk, S., 1966. Characteristics of some solonetzic soils in Glacial Lake Edmonton basin of Alberta, II. Mineralogy. *J. Soil Sci.*, 17 (1): 8 pp.

Bayrock, L.A. and Hughes, G.M., 1962. Surficial geology of the Edmonton District, Alberta. *Res. Counc. Alta. Prelim. Rep.*, 62-6: 40 pp.

Bowser, W.E., Kjearsgaard, A.A., Peters, T.W. and Wells, R.E., 1962. Soil survey of Edmonton sheet (83-H). *Alta. Soil Surv. Rep.*, 21: 66 pp.

Cairns, R.R. and Bowser, W.E., 1969. Solonetzic soils and their management. *Canada Dept. Agric. Publ.*, 1391: 23 pp.

Canada Department of Agriculture, 1970. *System of Soil Classification for Canada.* Canada Dept. of Agriculture, Queen's Printer, Ottawa, 249 pp.

City of Edmonton Planning Department, 1971. *Mill Woods, a Development Concept Report Prepared on Behalf of the Civic Administration*, 56 pp.

Hardy, R.M. and Associates, 1971. *Edmonton Public Schools Foundation Investigation, Grace Martin Elementary School, Mill Woods Development*, 14 pp.

Klingebiel, A., 1966. Cost and returns of soil surveys. *Soil Conserv.*, 32 (1): 3–6.

Pawluk, S. and Bayrock, L.A., 1969. Some characteristics and physical properties of Alberta tills. *Res. Counc. Alta. Bull.*, 26: 72 pp.

Swenson, E.G., 1971. Concrete in sulphate environments. *Canadian Building Digest, Division of Building Research, National Research Council of Canada*, 136: 4 pp.

United States Bureau of Reclamation, 1966. *Concrete Manual.* United States Department Interior, Bureau of Reclamation, 7th ed., 642 pp.

Geoderma, 10 (1973) 47–65

APPLICATION OF SOIL AND INTERPRETIVE MAPS TO NON-AGRICULTURAL LAND USE IN THE NETHERLANDS

G.J.W. WESTERVELD and J.A. VAN DEN HURK

Soil Survey Institute, Wageningen (The Netherlands)

(Accepted for publication October 2, 1973)

ABSTRACT

Westerveld, G.J.W. and Van den Hurk, J.A., 1973. Application of soil and interpretive maps to non-agricultural land use in The Netherlands. *Geoderma*, 10: 47–65.

The number of applications of soil maps and interpretive maps in the urban sector is gradually increasing in The Netherlands, especially in the western part of the country. This so-called "Randstad Holland" (Rim City Holland) is one of the most densely populated areas in the world.

In this conurbation soils predominantly used for agriculture or horticulture in the past are increasingly being claimed for residential, industrial, transport and recreational purposes. Yet for human welfare it seems imperative that agricultural land use be maintained as far as possible since it provides green belts separating the urban and industrial areas.

However, if land use is changed – the decision being usually determined by non-pedological factors such as geographical location – drastic soil improvement measures are usually needed to make the soils suitable for non-agricultural purposes. Interpretive maps indicating, for example, the nature and costs of these measures for the various soils, play an important part as is shown in a number of examples.

In The Netherlands the preparation of such maps is being increasingly preceded by multidisciplinary studies by working parties.

INTRODUCTION

The western part of The Netherlands is characterized by a high degree of urbanization with a high density of population (450–> 1,000/km^2, Fig.2A) and intensive agricultural land use. The area available for recreation is less than 40 m^2 per person. Except along the coast there is a virtual absence of forests and areas of natural beauty.

Especially after the World War II a number of major changes in land use took place because of expansion of towns and industrial sites and increased leisure time.

These changes gave rise to the following problems.

(1) Changes in land use are invariably at the expense of agricultural use of land.

(2) As regards soil conditions, a large area of the soils is unsuitable for urbanization, industrialization, road construction and intensive recreation. The soils usually have to be made suitable through drastic measures.

Hence the work of Dutch soil scientists differs markedly from that of their colleagues in many other countries. The latter mainly restrict themselves to describing soil conditions and soil suitability. In The Netherlands, soil surveys in or near urban areas also attempt to predict the potentials of soils for alternative land uses and the measures and costs involved in making them suitable.

This requires multidisciplinary research involving not only soil experts (soil scientists, geologists, hydrologists, soil engineers) but also town planners, landscape planners, etc.

The results are given in soil maps and interpretive maps in which, in addition to soil conditions and actual soil suitability, potential suitability is indicated with the investments thought necessary for soil improvements.

In this article the use of interpretive maps for the expansion of towns and for large recreation projects in the western part of The Netherlands, the so-called "Randstad Holland" (Rim City Holland), is illustrated by means of some examples. "Randstad Holland" is the area enclosed by the conurbations of Amsterdam, Haarlem, Leiden, The Hague, Rotterdam, Gouda and Utrecht (Fig.1 and 2).

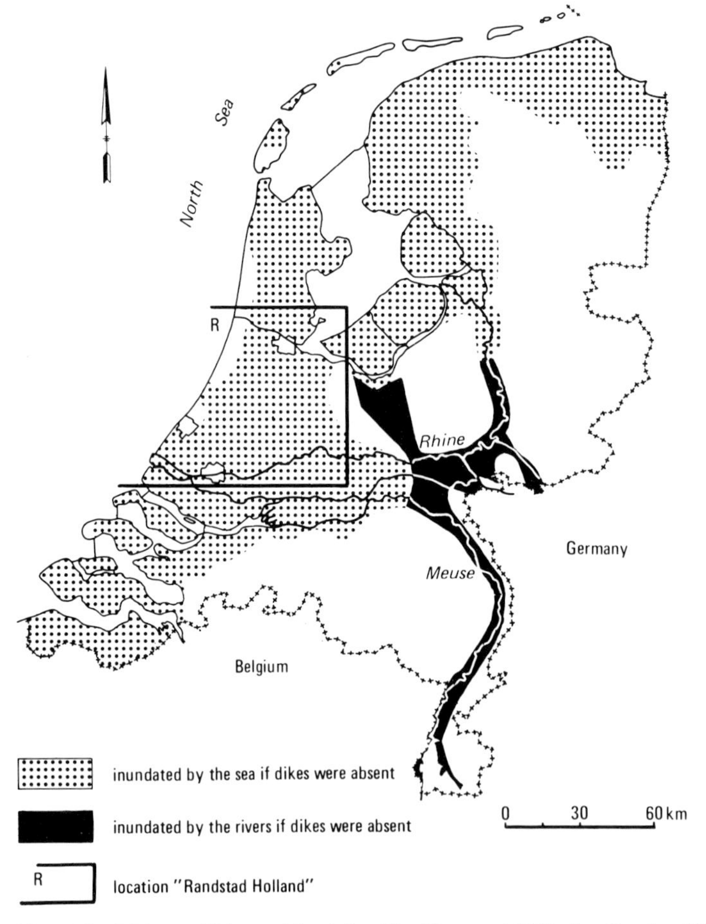

Fig.1. The Rhine and Meuse delta (after Van Heesen, 1970). *R* = Randstad Holland (see Fig.2).

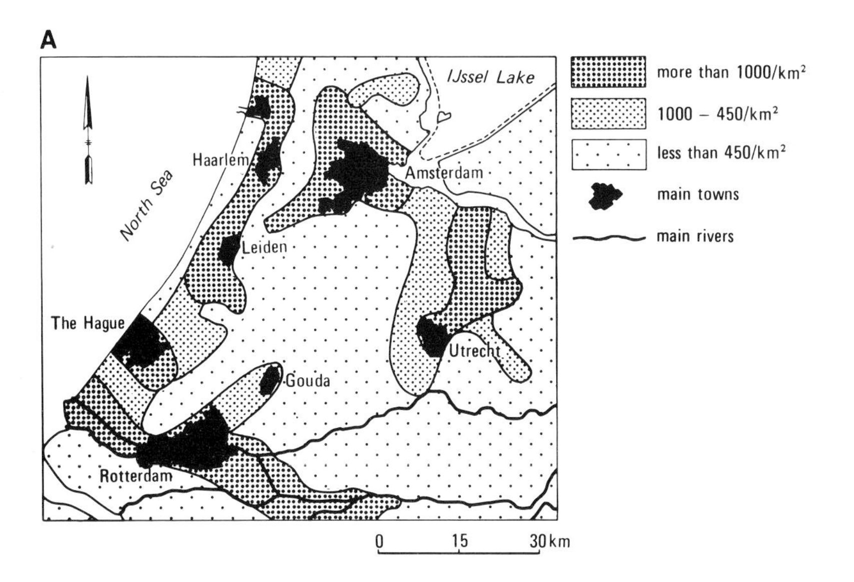

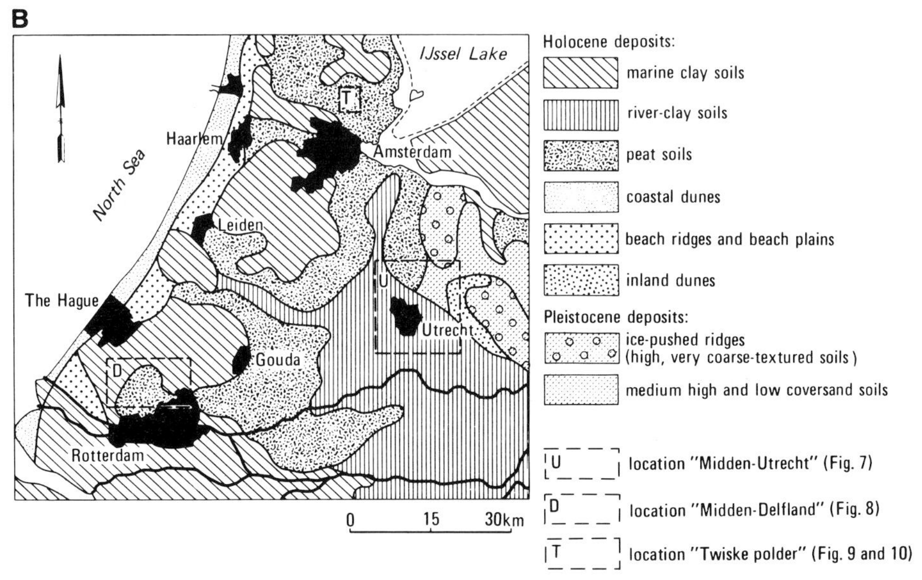

Fig.2. A. Density of population in "Randstad Holland".
B. General soil map of "Randstad Holland".

SOIL CONDITIONS AND HABITATION

"Randstad Holland" lies in the Rhine and Meuse delta, i.e., originally a coastal marsh (Fig.1). The soils consist mainly of layers of more or less unripened alluvial clay and peat, alternating with layers of firm, fine sand. This Holocene sediment changes at a depth of 10–15 m below surface to Pleistocene sand with a good bearing capacity.

The clay and peat soils, situated mostly in polders*, on the whole have winter groundwater levels within 0.40 m below the surface. Dry sands with good bearing capacity are found only in a narrow strip of coastal dunes and beach ridges along the North Sea and in the ice-pushed ridges and coversand deposits to the east of a line drawn between Amsterdam and Utrecht (Fig.2B).

In the past soil conditions have strongly defined the sites and forms of settlements and agricultural land use. This is apparent from the concentration of ancient settlements on higher and drier areas with good bearing capacity, such as the beach-ridge soils along the sea (Fig.3) and the natural levee soils along the rivers (Fig.4).

Fig.3. Settlement on high and dry beach-ridge soils with a good bearing capacity. On both sides are lower and wetter beach-plain soils with less bearing capacity. Photograph by courtesy of KLM Aerocarto B.V.

With increased population, soils less suitable for habitation had to be used. In Amsterdam as early as the Middle Ages buildings were erected on piles because of the low bearing capacity of the subsoil, and sand was brought in for reinforcement and elevation of the soil.

Nowadays soil improvements of a far more drastic nature are used in "Randstad Holland".

*A polder is a tract of lowland enclosed by dikes or dams; the water level in the polder ditches is artificially maintained by engineering works.

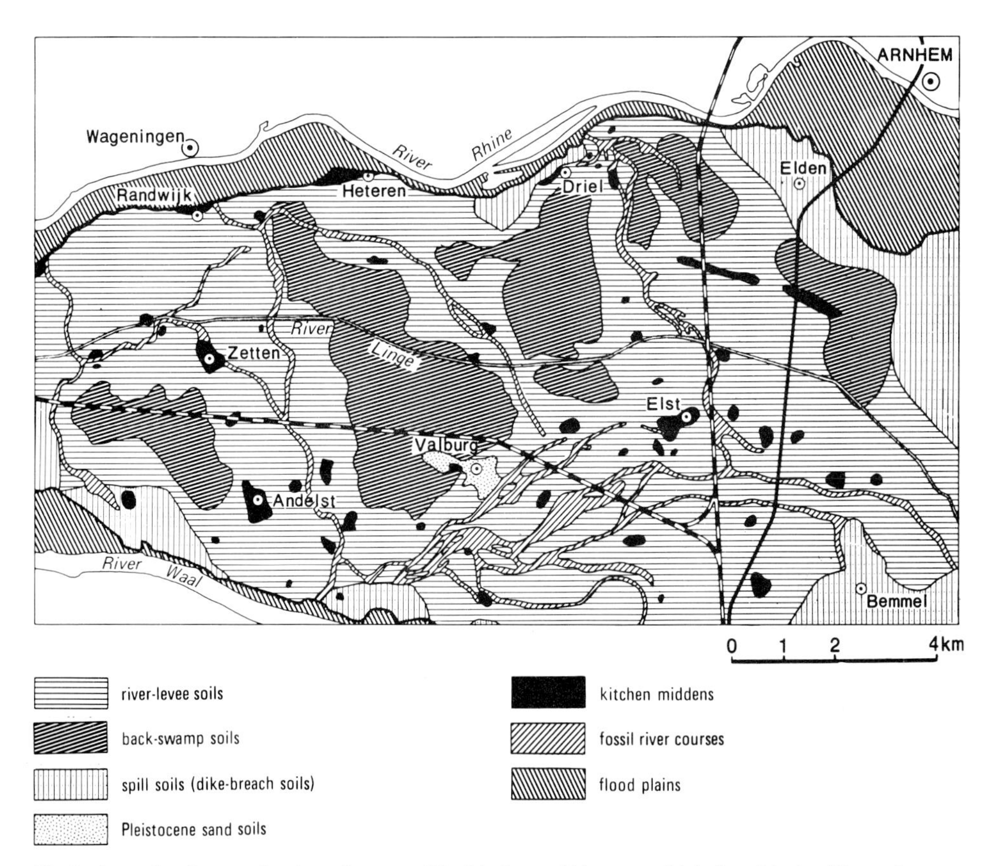

Fig.4. General soil map of a river-clay area. The kitchen middens on which the oldest settlements are centred, are situated on the high and dry river-levee soils with a good bearing capacity.

Technological developments have made it possible for buildings and roads to be planned on soils which are naturally unsuited for that purpose. Unfavourable soil conditions in general lead to increased costs, but these are rarely prohibitive when selecting new uses of land. For this purpose the geographical location of the land is far more important in The Netherlands.

TOWN EXPANSIONS SINCE 1945

During the last decades many towns have been expanded on naturally wet clay and peat soils with a low bearing capacity even at great depth. (Fig.2B). These soils were rendered suitable by addition of several metres of sand on top. This sand is largely extracted from underlying Pleistocene deposits. For this purpose the predominantly non-sandy Holocene overburden (thickness 10–15 m) first has to be removed. Sand extraction results in large areas of water which may be 40–50 m deep.

Building and road construction on raised areas offer many advantages from the techni-

Fig.5. New residential area built on low and wet soils with a low bearing capacity, which were elevated with 1–2 m of sand. The original landscape with its differences in soil, vegetation and relief has disappeared completely.

cal point of view. The areas have a good bearing capacity, deep groundwater levels and there is no uneven subsidence of the surface. However, building then takes place in surroundings which are unattractive for man and vegetation (Fig.5). Planning and maintenance of parks and playing fields present serious problems. The sand used on top is deficient in humus and clay and has to be improved by applying a thick layer of (highly) organic material in order to promote successful plantings.

Many new residential areas appear harsh and monotonous. With increased expectations of the quality of life, residents, town planners and landscape planners protest against this type of development.

As a result the possibilities of the existing landscape are now increasingly being considered in the planning and lay-out of new residential areas. A favourable housing and living environment is maintained wherever possible by preserving existing differences in relief, vegetation and soil conditions (Fig.6). In level landscapes even minor differences in relief help to break the monotony and to preserve existing vegetation patterns.

With regard to the planning of major expansions of towns within "Randstad Holland", factors other than soil conditions are decisive in most cases, and elevating with sand is often an inevitable measure. In many cases, however, it is not necessary to raise the surface of the whole site in question. For example, buildings of not more than two stories may be planned on clay and peat soils with Holocene sand layers within 1.00–2.50 m

Fig.6. New residential area built on high and dry soils with good bearing capacity, which were not elevated with sand. The original landscape with its differences in soil, vegetation and relief has been maintained to a large extent.

below surface, which have adequate bearing capacity for this purpose. In this case elevating with sand may be limited to those areas designated for other kinds of buildings. The location, shape and depth of the lake created by sand extraction may be designed to form a functional element in the greenbelt and to blend with its surroundings. The Holocene non-sandy overburden (spoil), produced from the lake may be spread over the lowest, wet areas designated for parks or playing fields. It may also be used as new topsoil for gardens and parks or playing fields in the areas elevated with sand.

Thus a residential area is created which has its own character. The newly planted vegetation has a favourable growing climate, whilst the old provides the necessary shelter and diversity.

Such a soil-orientated approach requires thorough pedological, geological soil engineering and hydrological preliminary studies. These should be carried out by multidisciplinary teams including town planners and landscape planners. The studies should cover not only inventory and design but also the possible alternative measures and their estimated costs.

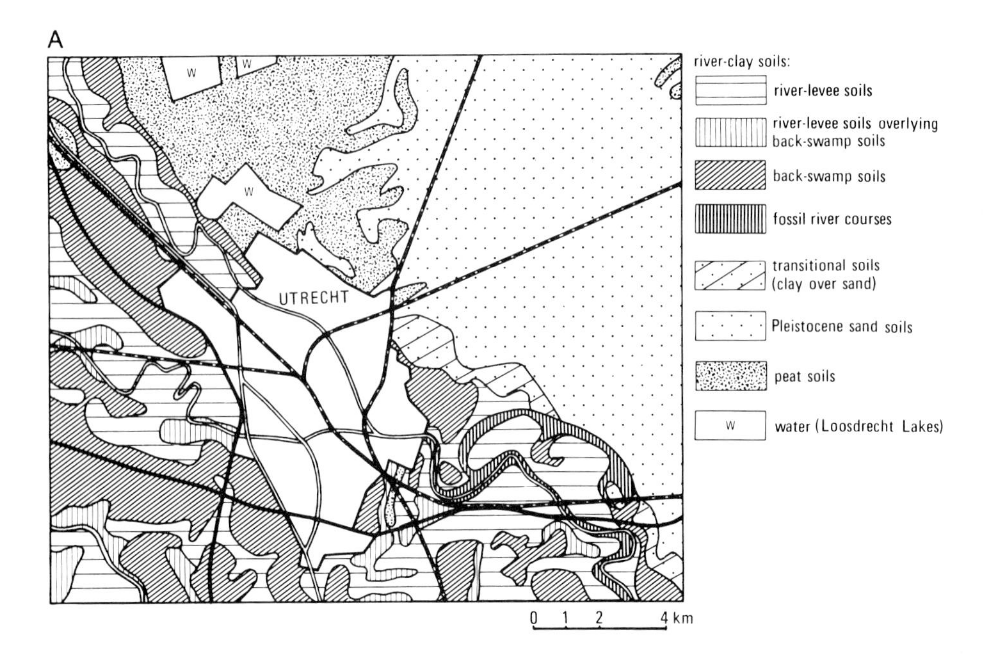

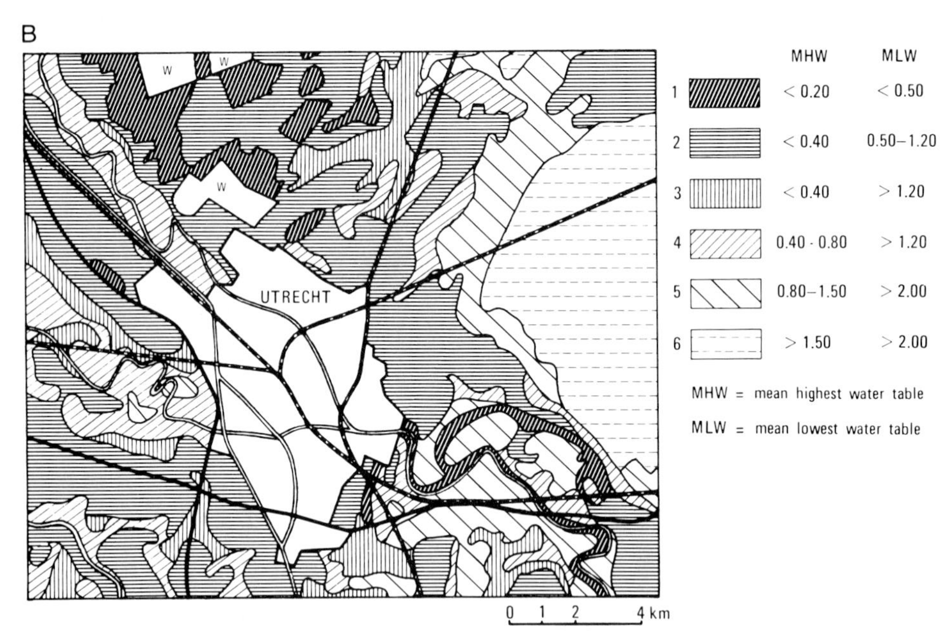

Fig.7. A. Soil map of the "Midden-Utrecht" area; for location see Fig.2B. (Original map, scale 1:50,000, according to Anonymous, 1970a and De Kievit, 1970)
B. Map of water-table classes (water tables in metres below surface).
C. Depth of Pleistocene sand (in metres below the surface).
D. Soil suitability map for building sites, based upon the costs of making soils suitable for buildings. (The costs are based upon the 1968–1969 price level, i.e., Dfl.1,000 = approx. US $ 280.)

C

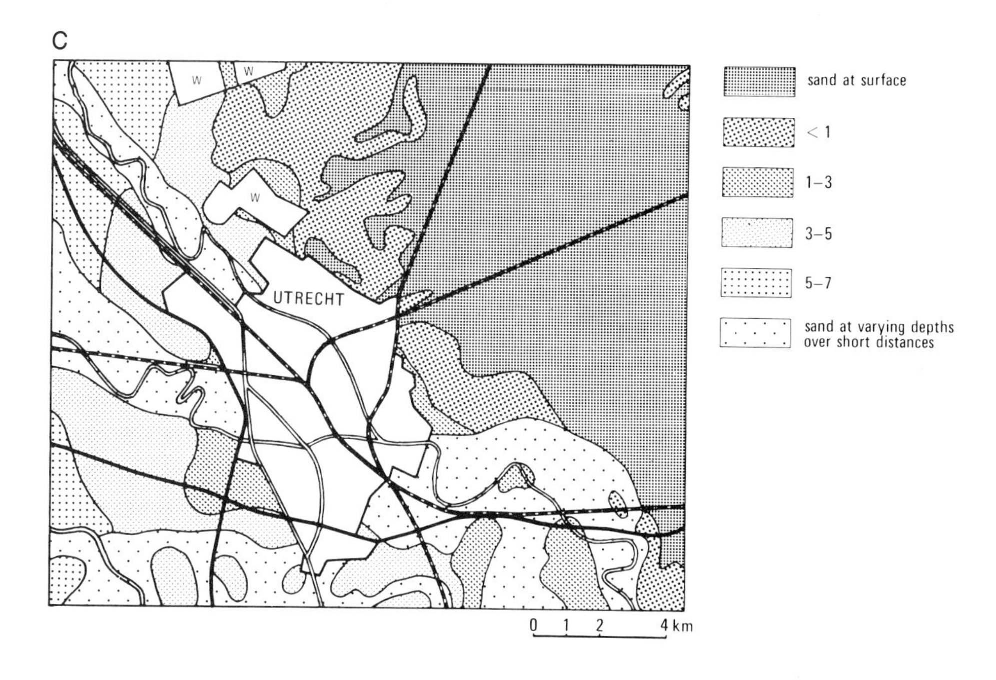
W
W
W
UTRECHT
sand at surface
< 1
1–3
3–5
5–7
sand at varying depths over short distances
0 1 2 4 km

D

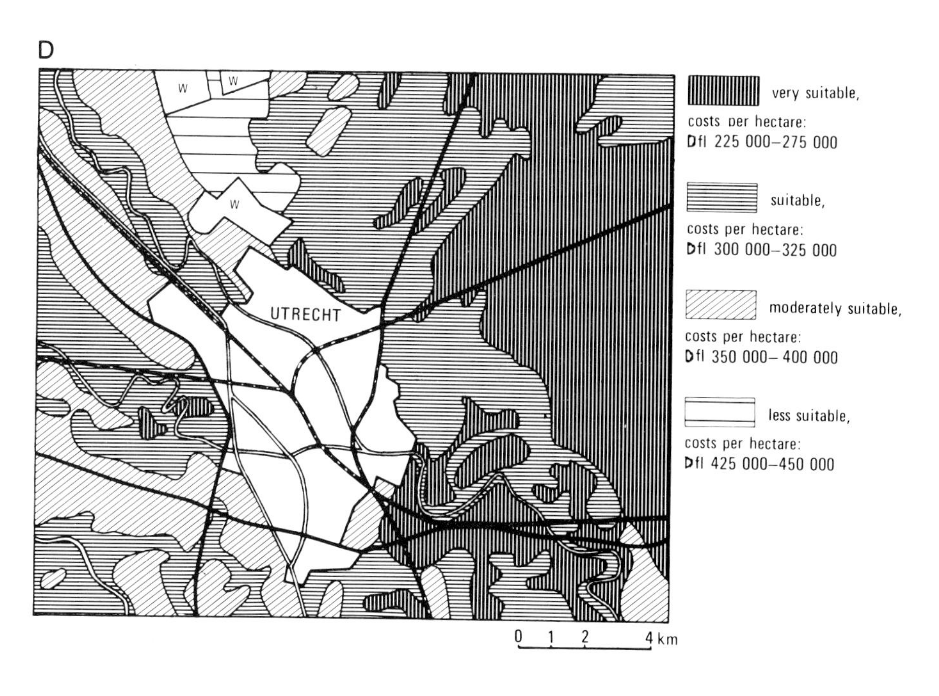
W
W
W
UTRECHT
very suitable,
costs per hectare:
Dfl 225 000–275 000
suitable,
costs per hectare:
Dfl 300 000–325 000
moderately suitable,
costs per hectare:
Dfl 350 000– 400 000
less suitable,
costs per hectare:
Dfl 425 000–450 000
0 1 2 4 km

The importance of a soil survey in this multidisciplinary and more soil-orientated approach is discussed on the basis of three examples. The location of the areas mentioned in the examples is given in Fig.2B.

APPLICATIONS OF SOIL SURVEYS IN "RANDSTAD HOLLAND"

*Soil suitability for building sites in the "Midden-Utrecht" area**

One of the aims of this study of the area around the fast-growing city of Utrecht was to indicate the possibilities for building on the basis of the costs of making the soils suitable.

Basic information was supplied by maps at a scale 1:50.000 of soil conditions down to approx. 1.20 m below the surface (Fig.7A), of the level and annual fluctuation of the groundwater (water-table classes) within approx. 1.50 m (Fig.7B), and of the depth of the Pleistocene sand (Fig.7C). In addition, information was available about the elevation, the water level in the ditches and the groundwater at greater depth.

The eastern part of the "Midden-Utrecht" area consists of high and dry Pleistocene sands which partly belong to a complex of ice-pushed ridges. Towards the west, these soils change into low and wet peat soils via coversand soils of low and medium height. Over the centuries the peat soils have been partly cut for fuel, resulting in a number of lakes (Loosdrecht Lakes). In the southern and westernmost areas, river-clay soils and transitional soils (river-clay over Pleistocene sand) are found.

The elevation ranges from 15 to 20 m above sea level on the ice-pushed ridges to approximately sea level in the peat and back-swamp soils. The groundwater level also varies strongly; it ranges from more than 5–10 m below the surface in the ice-pushed ridges, to less than 0.20 m in part of the peat soils (Fig.7B). In the clay and peat soils the depth of the Pleistocene sand ranges from less than 1 m below the surface in some peat soils to more than 9 m at some places in the back-swamp soils to the west (Fig.7C).

To assess soil suitability it is necessary to know the requirements of the relevant land use. With regard to building, these requirements mainly concern the groundwater level (drainage), bearing capacity (foundations) and relief. Having determined the requirements, the limitations of the soils are investigated. The extent of the limitations and the necessary measures and costs to offset them are then considered.

With regard to requirements, measures and costs the following standards have been used: groundwater must be more than 1 m below the surface for all soils; soils not meeting this requirement and for which the water table could not be lowered, have to be elevated with sand; the standard of elevating maintained was 1 m for sand and clay soils, and 1.50 m for peat soils where subsidence of the surface due to compression of the peat can be expected; a flat price per cubic metre of sand was adopted.

---

*The preliminary studies were carried out by the "Midden-Utrecht" multidisciplinary working party which included soil scientists. The results have been published in a number of reports (Anonymous, 1970a) and in a technical journal (De Kievit, 1970). The authors thank the members of this working party for information used in this article.

From the soil map and the water-table map (Fig.7A,B) and the information on the groundwater at greater depth, a decision can be made as to which of the soils should be totally or partially elevated. For all soils with water-table classes 3–6 and the majority of sand, transitional and river-levee soils with water-table classes 1 and 2, elevation was found to be unnecessary. The standard was met by lowering the groundwater level. In a small part of the soils having water-table classes 1 and 2, the high groundwater level is caused by underground water supply (seepage) from the ice-pushed ridges. Here the standard can only be met by elevating the soil with sand. Elevating is also required for back-swamp and peat soils having water-table classes 1 and 2.

A similar approach has been used to determine the costs of the foundations of the buildings and of the construction of roads and sewers. A decisive factor in the variations in cost is the depth to the underlying Pleistocene sand with adequate bearing capacity. This may be inferred from the map showing the depth to the sand (Fig.7C). Using this map and the soil map it was possible to estimate, for each mapping unit, the foundation depth and the costs of foundations and of road and sewer construction.

On the basis of the total costs per mapping unit, the soils have been divided into four classes (Fig.7D). In addition to the suitability of the soils, the investments necessary to make the soil suitable for buildings are indicated. For the very suitable soils these are Dfl. 225,000–275,000 per hectare and for the less suitable soils Dfl. 425,000–450,000 per hectare. The difference in cost between two classes is Dfl. 60,000–65,000 per hectare. The costs are based upon the 1968–1969 price level. The figures have no absolute value, but they constitute a significant difference in construction costs.

Comparing the soil suitability map for building sites (Fig.7D) with the three basic maps (Fig.7A–C), it appears that dry sand soils offer the widest range of possibilities and that the costs of making all wet back-swamp soils and peat soils suitable for building are high, due to the need for elevation of the soil. For soils with the Pleistocene sand deeper than 3m below surface, these costs are higher still due to the deeper foundations required.

Suitability of non-sandy spoil material from the Upper Holocene layers of sand excavations in the "Midden-Delfland" area

The survey carried out in the future recreational area known as "Midden-Delfland" has been described earlier in this journal (Haans and Westerveld, 1970). An aspect which is important to other recreational areas as well will be discussed below.

Attention has already been drawn to the fact that, following sand extraction from the deeper Pleistocene sediments for expansion of towns and road construction, the predominantly non-sandy Holocene overburden or spoil can often be utilized to improve soils designated for parks or playing fields.

For this method it is essential that the soil and geological surveys should not be limited to depth and quality of the underlying Pleistocene sand but that the nature and suitability of the overlying Holocene deposits should also be assessed. Location of sand extraction may also be determined by the suitability of these deposits.

A

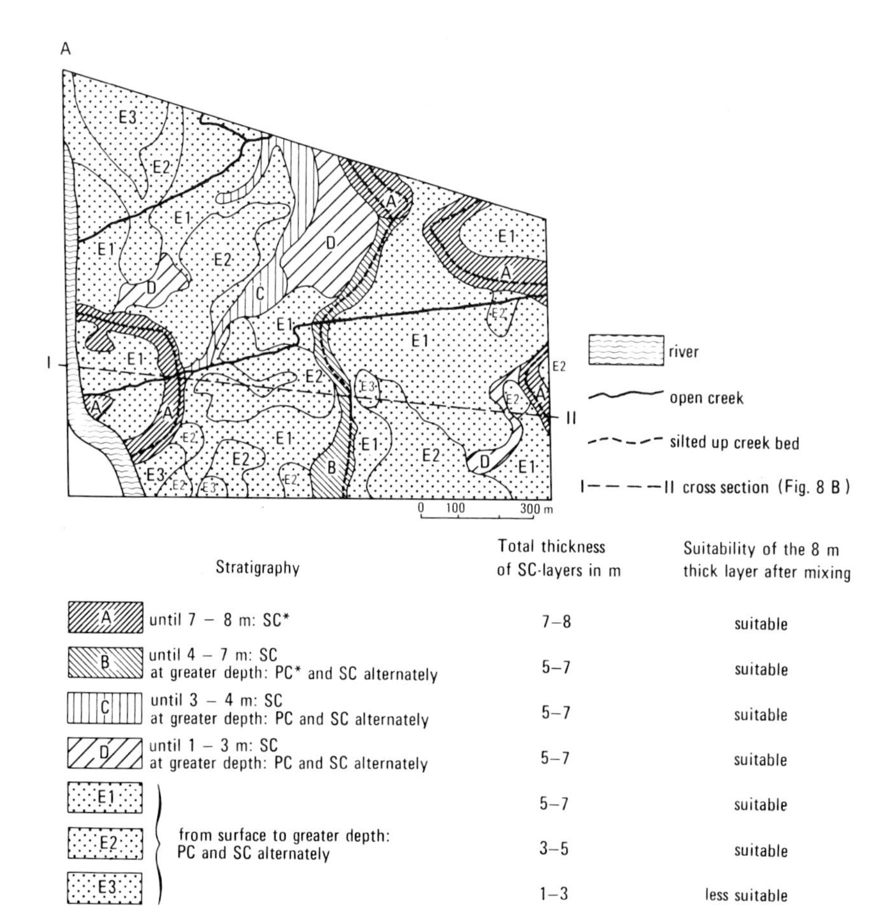

Stratigraphy	Total thickness of SC-layers in m	Suitability of the 8 m thick layer after mixing
A until 7 – 8 m: SC*	7–8	suitable
B until 4 – 7 m: SC at greater depth: PC* and SC alternately	5–7	suitable
C until 3 – 4 m: SC at greater depth: PC and SC alternately	5–7	suitable
D until 1 – 3 m: SC at greater depth: PC and SC alternately	5–7	suitable
E1 from surface to greater depth: PC and SC alternately	5–7	suitable
E2 from surface to greater depth: PC and SC alternately	3–5	suitable
E3 from surface to greater depth: PC and SC alternately	1–3	less suitable

* SC = sandy and or clayey material, calcareous   PC = peaty and/or peaty material, noncalcareous

B

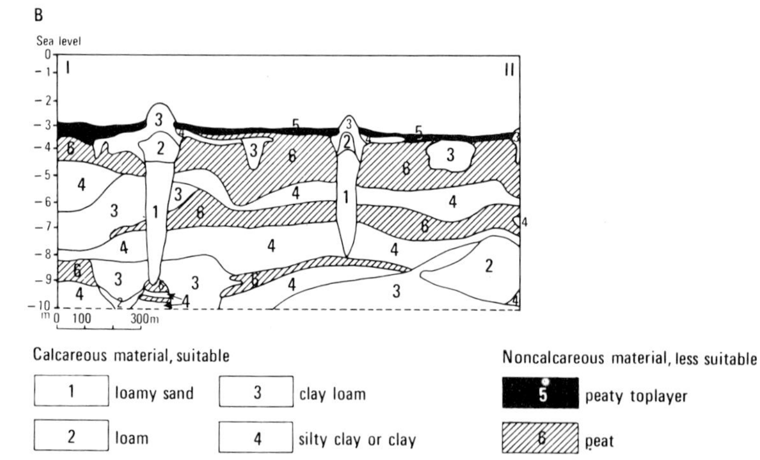

Fig.8. A. The "Midden-Delfland" area; for location see Fig.2B. Map showing suitability of material dredged from the 8-m thick Holocene layer, when used for elevating less suitable soils in places where parks or forests are planned.

B. Cross-section of the area shown in A (after Buitenhuis et al., 1971).

Such a survey was carried out in the future recreational area of "Midden-Delfland" (Fig.2B and 8). This area, surrounded on all sides by (future) residential and industrial sites, consists mainly of low and wet clay-on-peat soils of medium quality. These soils offer few possibilities for residential, sports and recreational purposes. They will have to be drastically improved by drainage and elevation. In addition, millions of cubic metres of sand will be needed for the lay-out of roads and of bordering residential and industrial areas. The sand must be taken mainly from the Pleistocene deposits which are found at a depth of 10–15 m below the surface and are covered by clay, peat and fine sand (Haans and Westerveld, 1970).

The findings of the survey on composition and suitability of the Holocene deposit are shown in a map and a number of cross-sections.

A section of the map with a cross-section are given in Fig.8 indicating the thickness of suitable and unsuitable layers to a depth of approx. 8 m below surface. The following two classes are distinguished: (a) material suitable as spoil to raise less suitable soils designated as parks or playing fields: sandy and/or clayey calcareous material; (b) material less suitable as spoil: peat and peaty material which is noncalcareous.

This is a rough classification because of lack of detail in the survey. Calcareous silty clay and clay turn out to be less suitable as top dressing for grass-covered playgrounds and playing fields than loam and loamy sand because of the high clay content. However, the clayey material is useful for areas to be planted with trees.

The method of spoil transport (piping) leads to mixing of suitable and unsuitable material. Separating suitable from less-suitable material is possible only if layers have considerable thickness. However, the advantage of mixing is that, if there is sufficient calcareous (suitable) material present in the spoil material, the unfavourable properties of the lime-deficient material will largely disappear. In view of the high $CaCO_3$ content of the calcareous material a satisfactory result may be achieved if suitable material exceeds 40–50%.

This means that the material from the upper 8 m of the mapping units A–E2 is still suitable as topsoil after mixing. E3 material is less suitable after mixing (Fig.8A,B).

Soil suitability for deciduous forest and grass-covered playgrounds in the "Twiskepolder" area

This area, comprising approx. 650 ha, is one of the lowest parts of the large lowland peat area north of Amsterdam (Fig.9). Here the peat layer originally had a thickness of 3–4 m. The underlying unripened clay layer has a thickness of 0.5–1.50 m. The clay gradually changes into fine Holocene sand. The underlying Pleistocene sand starts at 12–14 m below surface. Until World War II peat was cut unsystematically for use as fuel.

In 1943 the area was enclosed by dikes and drained. The southern part was reclaimed and is now being used as grassland, although the quality is poor. The remaining area continued to be peat marshland with irregularly shaped pools in which peat was forming and with strips of noncutover peat land.

At present sand extraction has formed a lake of approx. 65 ha with a depth of 30–40 m in the central area of the "Twiskepolder". Prior to sand extraction the peat and clay layers which were not useful were piped onto part of the neighbouring peat marsh. In this manner a spoil dump comprising approx. 80 ha was formed in the northern part of the area during the years 1965–1967 (Fig.9).

Once sand extraction is completed, the "Twiskepolder" will be made into a recreational area for the inhabitants of the urban agglomeration of Amsterdam. A provisional plan has been prepared for the future layout of the area in which the lake features as an important swimming and yachting centre. In addition, a large number of waterways are to wind their ways through the area. The remaining land is to be planted with trees to form a forest or, alternatively, to be made into grass-covered playgrounds and picnic areas. Part of the peat marsh is to be maintained as a nature reserve.

For the purposes of afforestation and laying out of recreational areas, a study was made of the soil conditions, soil suitability, necessary improvements and costs★. The results of this study for the northern spoil dump will be discussed below.

Fig.9. Northern part of the "Twiskepolder" area and environs; for location see Fig.2B. Photograph by courtesy of Aero-Camera of Rotterdam. *1* = northern spoil dump with temporary drainage system; *2* = artificial lake with suction-pump dredger; *3* = spoil dump in stage of development; *4* = peat marsh (planned nature reserve); *5* = peat soils in levelled and drained former peat dredgings, reclaimed since 1943 and used as grassland; *6* = dike and circular canal of the "Twiskepolder"; *7* = old cultivated peat soils used as grassland.

★This study was also carried out by a multidisciplinary working party which, in addition to both authors, included members of the Institute for Land and Water Management Research at Wageningen, Grontmij N.V. at De Bilt, and the Provincial Physical Planning Service at Haarlem. The results have been published in two reports (Van Wijk and Van den Hurk, 1971) and in a technical journal (Van Wijk, 1973).

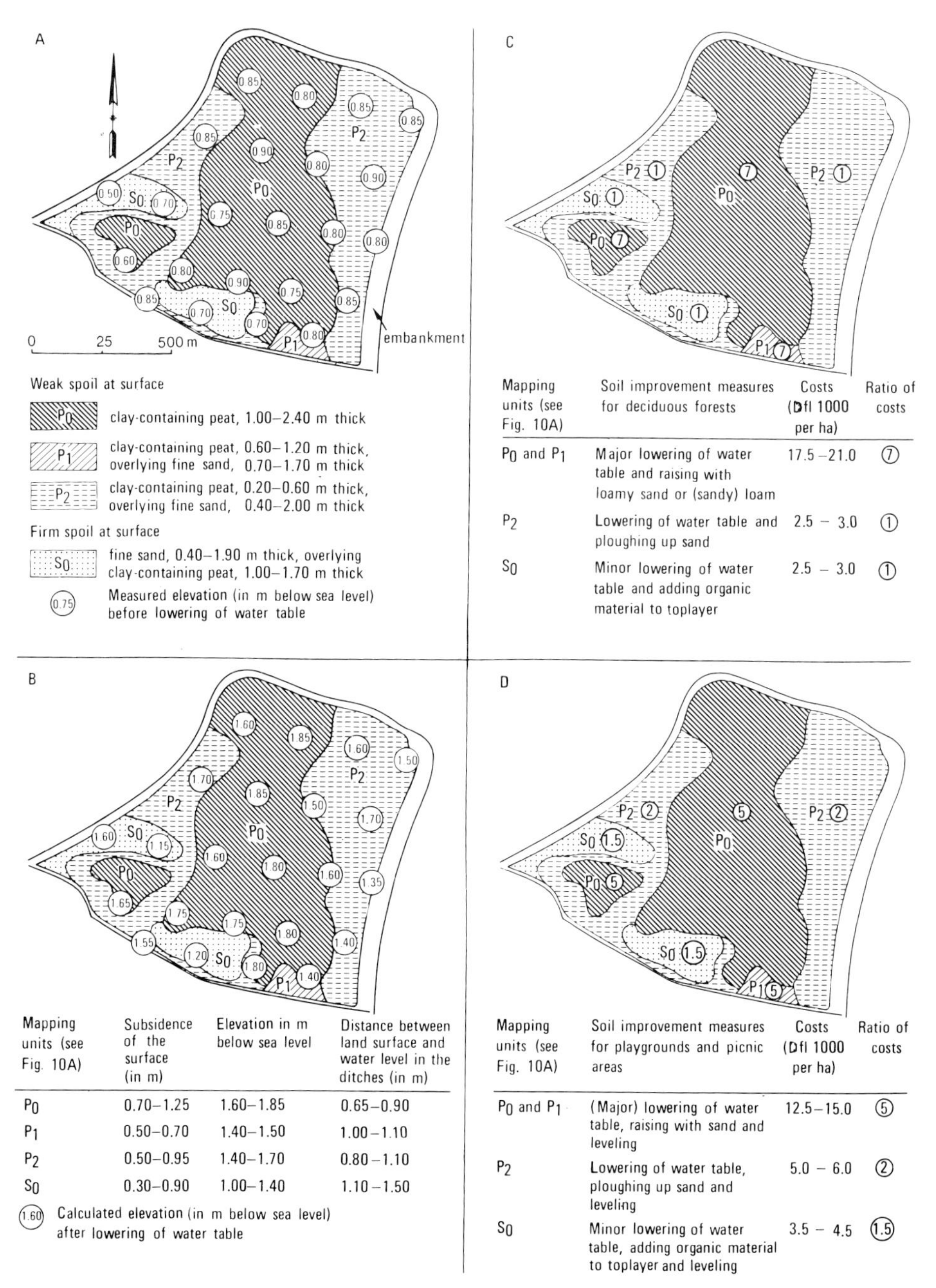

Fig.10. A. Soil map of the northern spoil dump of the "Twiskepolder" with elevation as measured before lowering of the water table. (Original map, scale 1:5,000 according to Van Wijk and Van den Hurk, 1971.)

B. Soil map of the area in A showing data as calculated after lowering of the water table to approx. 2.50 m below sea level.

C. Soil map containing information on the investments required per mapping unit to make soils suitable for deciduous trees. (The costs are based upon the 1970 price level, i.e., Dfl. 1,000 = approx. US $ 280.)

D. Soil map containing information on the investments required per mapping unit to make soils suitable for grass-covered playgrounds and picnic areas. (The costs are based upon the 1970 price level.)

As a result of the irregular surface caused by peat cutting, the thickness of the material piped onto the dump ranges from 0.80 to approx. 2.40 m with strongly varying soil conditions.

The soil consists of unripened (weak) spoil (Fig.10A, unit P_0) or weak spoil on fine sand (P_1 and P_2). In some parts of the dump the weak spoil is covered by fine sand (S_0). Except for these sandy parts the dump could be traversed on foot only during periods of frost in the first few years following piping. When a temporary drainage system was installed, the bearing capacity of the top layer increased and the surface began to subside as a result of consolidation of the unripened spoil. Prior to the lowering of the water table the surface was 0.50–0.90 m below sea level (Fig.10A).

In order to determine the necessary soil improvement measures, technical requirements were laid down for the soils in the case of each of planned deciduous forests, grass-covered playgrounds and picnic areas (Table I).

Afforestation had to be restricted to deciduous trees because soil conditions and climate in this area offered few possibilities for coniferous trees. With regard to the playgrounds and picnic areas, full account was taken of intensive use of these grass-covered areas.

For the planned uses, lowering of the water table to approx. 2.50 m below sea level is a major requirement. This will lead to subsidence of the surface through consolidation of the unripened peaty spoil.

Based on the findings of soil survey, the subsidence of the surface to be expected for the various soils and the future height of the surface following lowering of the water table, as well as the distance between land surface and water level in the ditches, have been calculated (Fig.10B).

It is clear from comparison of the maps in Fig.10A and B that the surface will not subside evenly. The strongest subsidences (0.70–1.25 m) will occur in soils having the thickest layer of peaty spoil (P_0). The difference between the eventual height of surface and the water level in the ditches will be the least in these soils, i.e. approx. 0.65–0.90 m. The surface of the sand soils (S_0) will sink the least (0.30–0.90 m). Here the difference between the land surface and the water level in the ditches will be the greatest (1.10–1.50 m).

In a manner similar to the one used in the "Midden-Utrecht" area, the necessary soil improvement measures are recommended and the costs estimated to make the soils suitable for the designated deciduous forest, grass-covered playgrounds and picnic areas. The costs are given in ratios in the maps in Fig.10C and D. Calculated at the 1970 price level, one cost unit amounts to Dfl. 2,500–3,000 per hectare.

Comparison shows that the soil improvement costs for deciduous forest on soils with the thickest layer of peaty spoil (P_0) are seven times that of sand soils (S_0). As is the case with the "Midden-Utrecht" survey, these figures have no absolute monetary value.

TABLE I

Technical standards to be met by the soils in the "Twiskepolder", so as to be suitable for deciduous forests and grass-covered playgrounds and picnic areas (after Van Wijk and Van den Hurk, 1971)

Land use	Soil conditions	Depth of water table	Distance between land surface and water level in ditches	Other requirements
Deciduous forests (including *Acer, Alnus, Fagus, Fraxinus, Populus, Salix* and *Quercus*)	topsoil of clay-containing mineral material as well as: $>$ 0.15 m thick for *Acer, Alnus, Fraxinus, Populus* and *Salix*; $>$ 0.50 - 0.60 m thick for *Fagus* and *Quercus*	$>$ 0.20 m for *Alnus, Populus* and *Salix* $>$ 0.30 m for *Acer* and *Fraxinus* $>$ 0.40 m for *Fagus* and *Quercus*	$>$ 1.00 m	pH-KC1$>$ 4.5–5.0
Grass-covered playgrounds and picnic areas (used intensively)	topsoil of good permeability and containing: $<$ 5% organic matter $<$ 5% clay (particles $<$2 μm) $<$ 10% clay + silt (particles $<$ 50 μm)	$>$ 0.30 m	$>$ 1.00 m	some micro-relief permitted

DISCUSSION

In spite of predominantly unfavourable soil conditions for town planning, industrial sites and road construction in the Rhine and Meuse delta area, known as Randstad Holland, this small area of The Netherlands holds more than a third (approx. 5.5 million inhabitants) of the Dutch population (Fig.1 and 2).

The increasing expansion of the urban areas is leading to a growing lack of suitable soils. Hence, drastic improvement measures for less suitable soils have become essential. So far improvement usually consists of elevating the less suitable soils by adding one or more metres of sand, depending on the kind of building to be erected. This leads to high establishment costs for urban and industrial buildings and, in addition, to uniform and unattractive housing and living environments because original differences in soil and landscape are largely eliminated (Fig.5).

For this reason, soil conditions are playing an increasingly greater role in planning of recent expansions. Soils must be considered in addition to the social, economic, planning and administration factors. In addition to mapping of actual and sometimes valuable differences in soil conditions and soil suitability, Dutch soil scientists contribute mainly in recommending measures and estimating costs required to make less suitable soils fit for a particular purpose.

Soil maps and interpretive maps showing the differences in soil conditions and soil suitability based upon investment costs to make soils suitable, can play an important part in the final planning. General maps are most useful when available at the time of formulating land-use and land-development patterns. However, these general maps are not a substitute for more detailed maps with scales of 1:10,000 and larger when plans are being executed. It is not a matter of using either general or detailed maps but of using both, one in the planning and the other in the execution phase.

If the maps contain information relevant to the land-use designations, they make an important contribution towards cost reduction and an attractive environment.

In order to achieve optimum location and planning of urban areas it is desirable that surveys be carried out by multidisciplinary teams.

REFERENCES

Anonymous, 1970a. *Beleidsrapport in Opdracht van de Kring Midden-Utrecht, Globale Visie. Hoofdlijnen voor de Inrichting van de Ruimte in Midden-Utrecht.* Amsterdam, München, Wageningen, 165 pp.

Anonymous, 1970b. *Bodem en Planologie. Een Studie over de Betekenis van de Bodemkartering voor Niet-Agrarisch Bodemgebruik. Stedebouw Volkshuisvesting*, extra nummer, 60 pp.

Buitenhuis, A., Van den Hurk, J.A. and Westerveld, G.J.W., 1971. *Midden-Delfland. Bodemgesteldheid en Bodemgeschiktheid.* Rapport 818, Netherlands Soil Survey Institute, Wageningen, 113 pp.

De Kievit, J.L., 1970. Een classificatiesysteem voor een bodemgeschiktheidsbeoordeling van gronden voor een bepaalde vorm van bodemgebruik. *De Ingenieur*, 45: 159–168.

Haans, J.C.F.M. and Westerveld, G.J.W., 1970. The application of soil survey in the Netherlands. *Geoderma*, 4: 279–311.

Van Heesen, H.C., 1970. Presentation of the seasonal fluctuation of the water table on soil maps. *Geoderma*, 4: 257–279.

Van Wijk, A.L.M., 1973. Een methode voor de evaluatie van de geschiktheid van gronden voor de inrichting van recreatieprojecten. *Recreatievoorzieningen*, 5: 16–23.

Van Wijk, A.L.M. and Van den Hurk, J.A., 1971. *Twiskepolder, Noordelijk Speciedepot, Geschiktheid voor Speel- en Ligweiden en Bos.* Werkgroep Bodem en Water Twiskepolder, Rapport 1, Institute for Land and Water Management Research and Netherlands Soil Survey Institute, Wageningen, 51 pp.

Geoderma, 10 (1973) 67–74

SOIL SURVEYS – THEIR VALUE AND USE TO COMMUNITIES IN MASSACHUSETTS

STEPHEN J. ZAYACH

Soil Conservation Service, U.S. Department of Agriculture, Amherst, Mass. (U.S.A.)

(Accepted for publication August 24, 1973)

ABSTRACT

Zayach, S.J., 1973. Soil surveys – their value and use to communities in Massachusetts. *Geoderma*, 10: 67–74.

For many decades, the study and mapping of soils in Massachusetts has primarily been related to farming with little or no attention given to other uses. About 1960 important changes in the use of soil surveys occurred in the state. Planners and communities drastically increased their demand for soil surveys and, particularly, for nonagricultural soil interpretations. Towns and cities have contributed $1/2 million since 1962 to accelerate soil surveys and have received special soil reports for operational planning. They have saved millions of dollars through the use of soil survey information. A number of examples are cited indicating the use and value of soil data to planners and communities.

INTRODUCTION

The kinds, characteristics, distribution, and extent of different kinds of soils are determined by means of soil surveys. The nature and distribution patterns in turn are important to the usefulness of soils for many purposes. This is becoming increasingly evident in the United States as the population expands and requires greater amounts of land for nonagricultural uses (Simonson, 1966). Investments per unit area of land are frequently high under such use, and mistakes are costly. Mistakes can often be avoided and more intelligent decisions on the use of land can be made if the nature and distribution of the soils are known. Consequently, a number of examples of how soil survey information was used in Massachusetts are described in this article.

HISTORY OF SOIL SURVEYS IN MASSACHUSETTS

One of the first soil surveys in the United States was made along the Connecticut River in Massachusetts and Connecticut (Dorsey and Bonsteel, 1899). The purpose was to provide information about the soils to improve farming practices. This continued to be the primary objective of soil surveys in Massachusetts for the next 60 years. Special attention was given to the properties of soils important to plant growth, moisture retention, tilth, and the like.

The emphasis on soil surveys has changed drastically in the state since 1960. About that time, planning consultants and town and city officials were searching for ways and means of improving the bases for decisions on the use of land, especially for nonagricultural uses (better known in the United States today as town and country planning). It soon was realized that soil properties as described in soil surveys could be used to predict the behavior of soils for uses other than farming, such as land use planning; residential, commercial, and industrial development; school sites; zoning and other land use controls; environmental quality planning; as well as farming, woodland, and wildlife.

Even though planning consultants and others asked for help from the Soil Conservation Service, they were not sure of the kinds of soil interpretations needed nor how the information should be presented and used. This led to the preparation and distribution of a publication on the subject. The publication was provided to communities, planning consultants, and others. It showed how soil survey data could be used in community planning. In preparing the report, the authors met with members of the Division of Planning (Massachusetts Department of Commerce), planning consultants, and sanitarians to determine the kinds of soil interpretations that would be useful to planners in making land use decisions.

The Massachusetts Department of Commerce, Division of Planning, published the report entitled *Soils Interpretation for Community Planning* in two volumes in 1963. Volume I, "Case study for the town of Hanover, Plymouth County, Massachusetts", deals with soils and their interpretation for various uses such as sewage effluent disposal, home sites, athletic fields, sources of sand and gravel, and development of wetlands for waterfowl (Zayach and Upham, 1963). The detailed soil maps of the town of Hanover have been converted to nine single-purpose interpretive maps for as many different uses. These interpretive maps indicate the suitability of soils for the various uses by three degrees of limitation – slight, moderate, and severe.

Volume II, "Effectuation of soils interpretation for the town of Hanover", demonstrates how soil data serve as a tool in planning and discusses their use in general planning and also in plan implementation (Thomas, 1963).

Subsequent to publication of *Soils Interpretation for Community Planning* and to many discussions throughout the Commonwealth of Massachusetts, there has been a great demand for soil survey information. To satisfy this demand, the Soil Conservation Service requested that communities share the costs for acceleration and completion of a soil survey. This approach has made it possible for communities to receive a special soil report for operational planning within a year or two after the soil survey was started.

To date, 130 of the 351 towns[★] and cities in the Commonwealth of Massachusetts have contributed funds toward the cost of soil surveys and received special soil reports. These reports contain copies of the detailed soil maps on aerial photobases, a general soil map, an airphoto index map showing the location of each detailed soil map, interpretive maps

[★]In Massachusetts, towns are local units of government for geographic areas of several thousand hectares (some thousands of acres).

for 10 or more different uses selected by the community, and descriptions of the soils, soil interpretations, and an explanation of how to use the report. The 130 communities have contributed $1/2 million to accelerate soil surveys.

VALUE AND USE OF SOIL SURVEYS TO COMMUNITIES

Millions of dollars have been saved by communities in Massachusetts by using soil survey information to select school sites, control subdivision developments, protect community water supplies, advise developers on sewage disposal systems, and in other ways. Several communities in the state indicated an average benefit–cost ratio of more than 110 to 1 in savings effected by using soil surveys to avoid errors in use of land. This information is based on data obtained in 1966 from 10 communities. These communities reported savings of $1,709,000 from limited use of soil surveys for a short period (1 to 3 years). The more than $110 benefit for each dollar expended for soil surveys is a conservative figure. When a community reported savings of $300,000 to $400,000, the lower figure was used in computing the benefit–cost ratio. Some of the communities reported benefits but did not have a dollar value attached to them. These would have increased the benefit–cost ratio.

The following examples are typical of the benefits derived from using soil surveys.

(1) The Planning Board of one community estimated savings of $500,000 in their school building program by using the appropriate interpretive map to select school sites. The savings resulted from having a complete inventory of suitable sites which permitted selection and purchase of the sites before land values increased. Additional savings resulted from purchasing the most suitable sites rather than waiting until site selection was limited.

(2) The Planning Board of another town rejected a development plan on some 50 acres on the recommendation of the Board of Health. The developer appealed to the state through the courts. The court decided in favor of the developer and ordered the town to grant the permit. The Planning Board appealed to a higher court, using soil survey data as the basis for their appeal. The higher court overruled the lower court and denied the permit.

The area was to have been developed for homes. However, there was no municipal sewer system, and soil survey data indicated a severe limitation on the use of septic tanks for sewage effluent disposal. The Planning Board indicated that this one incident saved the town more than the cost of the soil survey because it obviated the need to construct facilities that the town would have had to provide if the development had been allowed.

(3) The soils and related conditions of one town were such that future growth was being planned without a municipal sewer system – only onsite disposal systems were planned for new homes. Soil maps were used to zone the town to allow for adequate onsite sewage disposal. This eliminated the town's need for a $200,000 sewer system which would have been required by a hit-or-miss building program. Planning consultants also used soils information to adjust lot sizes. This postponed the need for an expensive sewer system until proper planning had been done for such a system.

(4) The U.S. Navy found that the soil maps of the Weymouth Naval Air Station saved them $25,000 to $30,000 in planning a runway extension because the soil data indicated areas where special problems could be anticipated. By making extensive borings and drillings only in the problem areas, the Navy greatly reduced the time and cost of foundation investigations.

(5) On the basis of soil surveys, another community determined that a municipal sewer system was not needed for a distance of 2 miles along one road because the soils had slight limitations for septic tank sewage effluent disposal. The total savings to the community were $105,600. The soil survey data were also used at a town meeting to show the need for funds for the extension of the municipal sewer system into an area of soils with severe limitations for onsite disposal systems, thereby forestalling onsite sewage disposal problems with attendant costs.

(6) A small community along the Atlantic coast estimated that the savings on the town sewer system will be in excess of $250,000 as a result of having soil survey information. The interpretive map for septic tank sewage effluent disposal has alerted the town to the fact that less than 1% of the 6,500 acres in the town had soils suitable for onsite sewage disposal. If 6,000 home-owners had established systems of their own, the disposal systems would not have functioned satisfactorily. This severe financial loss to home-owners was avoided because of advance knowledge that a municipal sewer system would be needed. The soil maps also indicated that a proposed 102-lot subdivision would endanger the town's water supply. A preliminary investigation and installation of a new well would have cost the town an estimated $2 million.

(7) One town at a special town meeting adopted and approved an amendment to its Zoning By-Laws that dealt with regulations for apartment buildings. Of the three items requiring approval for construction of apartments, one dealt with soil data. Unless connected to the municipal sewerage system, multifamily dwellings would not be allowed on soils classified as having a severe limitation for on-site sewage disposal. This decision was based on information obtained from the soil survey of Franklin County, Massachusetts (Mott and Fuller, 1967).

USE OF SOIL SURVEYS IN ENVIRONMENTAL QUALITY PLANNING

An ever-expanding population, particularly an affluent one, is likely to place demands on resources that may contribute to the deterioration of the quality of the environment. The environmental quality planning program, recently begun in Massachusetts by resource agencies, seeks to enhance or improve the quality of the environment (Isgur, 1972). It also seeks to promote the idea that the quality of the environment should be based on the carrying capacity of the land, water, and related resources, i.e., the capacity of the resources to support a given population consistent with maintaining and protecting an acceptable environmental quality.

The program emphasizes the physical characteristics and uses of natural resources, but it also is cognizant of social, economic, and political factors as potent forces that bear upon use of these resources. The program is envisioned as interdisciplinary in scope and having interagency inputs. The interagency expertise is offered to local citizen decisionmakers planning for a quality environment.

The purposes of the program are the following.

(1) To provide to communities or regions information about the quality, quantity, and distribution of their natural resources, based on environmental quality criteria established for use in Massachusetts. The environmental quality program indicates whether the carrying capacity of the natural resources is exceeded and, if so, the impact on the quality of the environment.

(2) To provide to decisionmakers alternatives in resource use and treatment for a given environmental quality that will meet the accepted objectives.

(3) To help planners in choosing from alternatives provided and in implementing the plan so as to maintain the quality of the environment or to improve it to the quality level of their choice.

An ad hoc committee, comprised of personnel from the Soil Conservation Service, was organized early in 1970. Their charge was to outline a methodology for evaluating an area's environmental quality on which to base the population-carrying-capacity of its natural resource base. The committee developed criteria and methodology to determine the environmental quality of a community. To test the suitability of the criteria and feasibility of the methodology, a preliminary study was carried out by the committee in a town in the eastern part of the state during the latter part of 1970. This town was selected because of the large amount of information available, such as soil surveys, natural resource inventory, town master plan, and inventory of potential reservoir sites.

Following the preliminary study, the criteria and methodology were reviewed, evaluated, and revised. The revised criteria and procedures were then reviewed with personnel from various state and other federal agencies. The ad hoc committee was expanded to include people from the above mentioned agencies, and they assisted in the revision of the criteria.

Another town in the eastern part of the state was selected as a prototype pilot project in 1971 to test the criteria and procedures. Local citizens participated in natural resource planning with the Natural Resource Technical Team of personnel from various state and federal agencies. Again the criteria and procedures were evaluated and revisions made based on the experience gained in the town.

In the fall of 1972, pilot projects were selected in five locations across the state – four in towns and one in a river basin. The latter was selected to test the program on a broader basis. A full-fledged field testing and evaluation of the criteria and procedures will be made in these pilot projects during 1974. The environmental quality planning program is expected to become operational in 1975.

Soil survey data are used in the evaluation and planning phase of the environmental quality program. The data are used to evaluate and rate the qualitative elements of the following resources: agricultural land, recreation land, woodland, wildlife land, urban land, and wildlife wetland. These data plus other information are then used to determine the environmental quality index, a value that indicates the environmental quality level compared to an established standard.

Soil data are also used in the final planning step-making decisions on land use and treatment. Proposed land uses are checked against soil data to verify that the proposed uses are on suitable soils. If the decisionmakers have decided on a certain land use on soils having moderate or severe limitations, soil data are used to indicate the location of areas having such limitations and the degree of conservation treatment needed to overcome limitations for the proposed use. Soil and other data are also used to help decisionmakers determine the people-carrying-capacity of the natural resource base.

USE OF SOIL SURVEYS FOR DETERMINING LOT SIZES

Carol Thomas, planning consultant, states: "City and town planners, or consultants, have to evaluate land use patterns and population densities and to develop comprehensive plans with local officials. Many tools and types of information are used in connection with this work of which soil data have become one of the most important. Soil information and interpretations provided by S.C.S. have provided a tool for determining population density, lot sizes, sewer, water and drainage system needs, location of house lots, and open space requirements.

"In the few years that the soil studies have been available to planners, their value has been demonstrated many times. Each community has realized savings which would pay for the study many times over. Proper lot size determination alone may obviate the need for costly utility estimates or prevent pollution of wells or water in basements of homes, thus saving millions of dollars. The study has become so important to us that we included it with the basic research of our planning program" (C. Thomas, personal communication, 1966).

For many years, planning consultants have requested that soil survey data be interpreted for lot sizes. These data, along with available soil interpretations for homesites, could be very useful when planners develop future land use maps for communities, especially in delineating areas for various densities of residential use. Experts from several disciplines met and discussed the possibilities of rating soils for this particular use. After several meetings, the soils of Massachusetts were rated for lot sizes for homes with and without municipal services. These data were included in a handbook for planners. The major factors considered in rating the soils for this use are: permeability; depth to slowly permeable soil horizons, bedrock, and seasonal high water table; surface stoniness and rockiness; soil slope; flooding by stream overflow; and hydrologic characteristics of soils as they effect surface water runoff.

A "Planners Handbook" has been published by the Massachusetts Federation of Planning Boards (Emilita, 1972). The purpose of this handbook is to assist Massachusetts planning boards by outlining the functions of planning boards as established by Chapters 40A and 41 of the Massachusetts General Laws. The appendices of the handbook contain a section entitled "Guide for lot size determination for single family dwellings". This section provides a guide for planners and planning boards engaged in comprehensive master planning activities (see Fig.1).

GUIDE FOR LOT SIZE DETERMINATION FOR SINGLE FAMILY DWELLINGS

er - extremely rocky
est - extremely stony
nst - nonstony
st - stony
vr - very rocky
vst - very stony
var - variant

Soils and Land Types	Soil Limitations for Home Sites with Cellars: No Municipal Services or Only Municipal Water [2]	Soil Limitations for Home Sites with Cellars: Municipal Water and Sewer	Lot Size Per Dwelling Unit in Square Feet [1]: No Municipal Services	Lot Size Per Dwelling Unit in Square Feet [1]: Municipal Water	Lot Size Per Dwelling Unit in Square Feet [1]: Municipal Water and Sewer	Lot Size Per Dwelling Unit in Square Feet [1]: Municipal Water, Sewer & Storm Drain
Warwick, 0-3% slopes	Slight [6]	Slight	40,000	20,000	10,000	10,000
" 3-8% "	"	"	"	"	15,000	"
" 8-15% "	Moderate [6]	Moderate	"	30,000	"	"
" 15-25% "	Severe (slope)	"	60,000	40,000	20,000	15,000
" 25-35% +"	" "	Severe (slope)		Not Feasible		
Watchaug nst, vst, & est, 0-3% slopes	Severe (wetness)	Moderate [3]	generally not feasible		15,000	10,000
" 3-15% "	" "	"	" " "		20,000	"
" 15-25% "	" "	"	" " "		25,000	15,000
Westminster (all map units)	Severe (bedrock)	Severe (bedrock)		Not Feasible		
Wethersfield nst, vst, & est, 0-3% slopes	Severe (hardpan-shallow)	Moderate [4]		Not Feasible	20,000	10,000
" 3-15% "	" "	"		" "	25,000	"
" 15-25% "	" "	"		" "	30,000	15,000
" est, 25-45% + "	" "	Severe (slope)		Not Feasible		
Whately, 0-3% slopes	Severe (wetness)	Severe (wetness)		Not Feasible		
Whitman (all map units)	Severe (wetness)	Severe (wetness)		Not Feasible		
Wilbraham (all map units)	Severe (wetness)	Severe (wetness)		Not Feasible		
Windsor nst, 0-3% slopes	Slight [6]	Slight	40,000	20,000	10,000	10,000
" 3-8% "	"	"	"	"	15,000	"
" 8-15% "	Moderate [6]	Moderate	"	30,000	"	"
" 15-25% "	Severe (slope)	"	60,000	40,000	20,000	15,000
" 25-35% +"	" "	Severe (slope)		Not Feasible		
Windsor vst, 0-3% slopes	Moderate [6]	Moderate	40,000	20,000	10,000	10,000
" 3-8% "	"	"	"	"	15,000	"
" 8-15% "	"	"	"	30,000	"	"
" 15-25% "	Severe (slope)	"	60,000	40,000	20,000	15,000
" 25-35% +"	" "	Severe (slope)		Not Feasible		
Windsor vr & er (all map units)	Severe (bedrock)	Severe (bedrock)		Not Feasible		
Winooski, 0-8% slopes	Severe (wetness)	Severe (flooding)		Not Feasible		
Woodbridge nst, vst, & est, 0-3% slopes	Severe (wetness)	Moderate [5]	Not feasible		20,000	10,000
" 3-15% "	" "	"	" "		25,000	"
" 15-25% "	" "	"	" "		30,000	15,000

FOOTNOTES FOR "GUIDE FOR LOT SIZE DETERMINATION"

[1] Approximate sizes indicated for single family dwelling in conventional subdivision – actual size is dependent on family size, number of bathrooms, on-site investigation, etc. When land is developed for home sites, it is assumed that the natural drainageways will be preserved to drain runoff water.

[2] The soil limitation ratings are based on use of septic tank sewage disposal system and an assumption the house lots are about one-half acre in size.

[3] Wetness problem due to seasonal high water table or excess seepage water for 3 to 5 months of the year. Foundation drains are needed to prevent wet cellars.

[4] May have wetness problem because of perched water table above hardpan during the wet parts of the year or prolonged periods of rain. Foundation drains are usually needed to prevent wet cellars.

[5] Wetness problem due to perched water table above hardpan for 3 to 5 months of the year. Foundation drains are needed to prevent wet cellars. Diversion ditches may be useful to help intercept surface and subsurface runoff water.

[6] The coarse textured substratum of these soils is so permeable that shallow wells may be contaminated when located close to septic tank disposal fields.

Fig.1. Last page from the "Guide for lot size determination for single family dwellings".

The Guide lists all the soils and land types that occur in the state. The map units, or groups of map units, are rated for homesites with cellars for two categories: "no municipal services or only municipal water" and "municipal water and sewer" available. Lot sizes in square feet per dwelling unit were established for each kind of soil. The lot sizes per dwelling unit are for the following four categories: no municipal services; municipal water; municipal water and sewer; municipal water, sewer, and storm drain.

REFERENCES

Dorsey, C.A. and Bonsteel, J.A., 1899. A soil survey in the Connecticut Valley. In: *Field Operations of the Division of Soils. U.S. Dept. Agric. Rept.*, 64: 125–140.

Emilita, D.J.S., 1972. *Planners Handbook*. Massachusetts Federation of Planning Boards.

Isgur, B., 1972. A suggested methodology for environmental quality planning of the natural resource base. In: *The Earth Around Us. Proc. 27th Ann. Meeting, Soil Conserv. Soc. Am.*, pp.166–167.

Mott, J.R. and Fuller, D.C., 1967. *Soil Survey of Franklin County, Massachusetts*. U.S. Dept. of Agriculture, Soil Conservation Service, in cooperation with Massachusetts Agricultural Experiment Station.

Simonson, R.W., 1966. Shifts in the usefulness of soil resources in the U.S.A. *Agriculture (Montreal)*, 23(3): 11–15.

Thomas, C., 1963. Effectuation of soils interpretation – Town of Hanover, Plymouth County, Massachusetts. In: *Soils Interpretation for Community Planning, II*. Massachusetts Department of Commerce, Division of Planning, and United States Department of Agriculture, Soil Conservation Service, pp.1–22.

Zayach, S.J. and Upham, C.W., 1963. Soils and their interpretations for various uses – Town of Hanover, Plymouth County, Massachusetts. In: *Soils Interpretation for Community Planning, I*. Massachusetts Department of Commerce, Division of Planning, and United States Department of Agriculture, Soil Conservation Service, pp.1–49.

Geoderma, 10 (1973) 75–86

SOIL SURVEY AND INTERPRETATION PROCEDURES IN MOUNTAINOUS WATERTON LAKES NATIONAL PARK, CANADA*

GERALD COEN

Agriculture Canada, Soil Research Institute, Soil Survey, The University of Alberta, Edmonton, Alta. (Canada)

(Accepted for publication August 24, 1973)

ABSTRACT

Coen, G., 1973. Soil survey and interpretation procedures in mountainous Waterton Lakes National Park, Canada. *Geoderma*, 10: 75–86.

Detailed soil surveys in mountainous areas are greatly facilitated by the use of aerial photographs but because of the large amount of photodistortion in areas of high relief accurate transfer to planimetric base maps is difficult. Because of limited geodetic control photo mosaics of sufficient accuracy for use as base maps were not available. Information was therefore published on enlargements of small-scale photography. It was felt that a photo base for the map presented the information in the most useful format. By designing the legend to limit the number of different mapping units, interpretations for several uses were made without loss of cartographic detail. Other interpretations can subsequently be made with relative ease. Thus, a strong interaction prevails between the design of a soil survey (methodology) and the ease with which interpretive information can be generated and presented.

INTRODUCTION

Until relatively recently in Canada soil surveys have been conducted for inventory purposes largely on a reconnaissance basis (approximately 1 inch to 1 or 2 miles) with little attempt at systematic interpretation of the generalized information for purposes other than agriculture. Local survey units have commonly provided interpretative assistance for many uses on a user request basis. However, the Agriculture Rehabilitation Development Act of 1961 and the subsequent establishment of the Canada Land Inventory Program provided the impetus for systematic interpretations for several non-agronomic uses. The resulting increased awareness of the applicability of soil survey information to recreation land use led to a request by the Natural and Historic Parks Branch in the spring of 1971 for a soil survey of the 200 Square mile (500 km^2) Waterton Lakes National Park. This was undertaken cooperatively by Environment Canada, Canadian Forestry Service, and Agriculture Canada, Soil Research Institute with financial assistance by the Department of Indian Affairs and Northern Development, Natural and Historic Parks Branch.

*Soil Research Institute Publication No. 470, and Alberta Institute of Pedology Publication No. T-73-4.

DEVELOPMENT OF THE SURVEY PROGRAM

Mapping separations were based on soil properties, as outlined by the *System of Soil Classification for Canada* (Canada Soil Survey Committee, 1970). To have made separations based on suitability for a given use would have required a resurvey when another use was identified (Smith, 1965). The method adopted avoids this problem but another step is required to identify (interpret) the soil areas with desirable attributes for specific uses. In the context of soil surveys the interpretative information must be considered the final product (Table I) and the ability to obtain it quickly and accurately from the generalized information influences the methodology of the entire survey. Further reference is made on pp. 79 and 83 to the interaction between the design of the survey and the ease of obtaining and presenting the interpretive information.

Field mapping was done on aerial photographs having a distance of 3 inches on the photograph representing approximately 1 mile on the ground surface at 4,200 ft. above M.S.L. The minimum size of area delineated was between 2 and 4 ha (5 and 10 acres). This level of cartographic detail was chosen in an effort to supply sufficient information for projected requirements of local park officials. On photographs of the selected scale the smallest desired area on the ground corresponds fairly well with the smallest field delineation in which a mapper can place a symbol. This is a convenient guide to prevent mappers from attempting to recognize too much detail. There are many problems associated with a relatively detailed soil mapping project in a mountainous area with relief of about 4,000 ft. (1,200 m) in less than a mile (1.6 km) (Fig.1).

This project was undertaken as a pilot study, both from the point of view of pedology and the point of view of the usefulness of soil surveys to facilitate park operations. This was the basis for selection of the field mapping scale. Before leaving the topic of mapping intensity, an ambiguity in the meaning of scale should be mentioned. The "scale" of presentation of the information may not coincide with the "scale" used in field work to record information, particularly when several disciplines are gathering resource information for later synthesis or correlation. It is desirable that the final scale of presentation be the same as the field mapping scale to prevent waste of time gathering unused information or bulky presentation of limited information. In multidisciplinary studies, where information from diverse disciplines is to be compared, all resource information should be collected at the same level of intensity for its most useful synthesis.

METHODOLOGY OF THE SURVEY

Basic principles

At the proposed mapping scale, and given the high relief and inaccessibility of the area, it was necessary that the soil delineations be chosen so as to correlate as strongly as possible with easily visible landscape features and still provide the separations necessary to meet the objectives of the survey. This should be true in most soil surveys but it is especially

TABLE I

Example of some of the "Interpretations of Soil Characteristics for Selected Park Uses" in Waterton Lakes National Park (Coen and Holland, 1973).

Map Units	Degree and Nature of Limitations for Recreation Uses									Degree and Nature of Limitations for Engineering Uses										Susceptibility to water erosion
	Playgrounds			Camp Areas			Paths & Trails			Septic Tank Fields				Bldgs. with Basements			Local Roads			
	slight	moderate	severe	slight	moderate	severe	slight	moderate	severe	slight	moderate	severe	pollution hazard	slight	moderate	severe	slight	moderate	severe	
1/AC, 1/AD		Stony	Slope CF[1] Moist		CF Stony	Moist		CF Stony		nil			Po		Stony		Stony			Low
1/DE, 1/EF, 1/F		Stony	Slope CF Moist		Slope Stony CF	Moist		Slope Stony CF			Slope		Po		Stony Slope			Slope		Low
1/FG, 1/G, 1/GH		Stony	Slope CF Moist		Stony CF	Slope Moist		Stony CF	Slope		Stony	Slope	Po		Stony	Slope			Slope	Moderate
15/AB		Flood Wet			Flood			Flood				W.T. Flood			Wet Str	W.T. Flood		Flood Str	Frost	Low
17/AC, 17/AD			Moist CF		CF	Moist	nil			nil			Po	nil			nil			Low
21/AC			CF Moist		CF	Moist		CF		nil			Po	Stony			Stony			Moderate
21/F			CF Moist		Cf	Moist Slope		Slope Cf				Slope	Po			Slope			Slope	High
31/AB			Wet Flood			Wet Flood			Wet Flood		Perm	W.T. Flood			Sh-Sw Str Frost	Wet, W.T. Flood		Sh-Sw Frost	Wet Flood Str	Low
50/AD, 50/CD		Moist	Stony CF Slope		CF Stony Moist			CF			Perm				Stony			Frost Str		Low
50/DE		Moist	Slope CF Stony		CF Stony Moist Slope			CF			Perm Slope				Stony Slope			Frost Str Slope		Low
50/EF		Moist	Slope CF Stony		CF Stony Moist	Slope		CF Slope			Perm	Slope			Stony	Slope		Frost Str	Slope	Low
50/FG, 50/G		Moist	Slope CF Stony		CF Stony Moist	Slope		CF	Slope		Perm	Slope			Stony	Slope		Frost Str	Slope	Moderate
100/AC		Slope	Text			Text			Text			Perm			Sh-Sw	Str		Sh-Sw Frost	Str	Low
100/DE			Text Slope		Slope	Text			Text		Slope	Perm			Sh-Sw Slope	Str		Sh-Sw Frost Slope	Str	Low
100/EF			Text Slope			Text Slope		Slope	Text			Perm Slope			Sh-Sw	Str Slope		Sh-Sw Frost	Str Slope	Moderate
100/FG, 100/G			Text Slope			Text Slope			Text Slope			Perm Slope			Sh-Sw	Str Slope		Sh-Sw Frost	Str Slope	Moderate
100/H			Text Slope			Text Slope			Text Slope			Perm Slope			Sh-Sw	Str Slope		Sh-Sw Frost	Str Slope	High

[1]Abbreviations as follows: CF = Coarse fragments; Stony = Surface stoniness; Moist = Useful moisture; Slope = % slope; Po = Pollution hazard; Flood = Flooding; Wet = Wetness (soil drainage); W.T. = Depth to seasonal water table; Str = Strength of soil as rated by Unified or AASHO; Frost = Susceptibility to frost heave; Perm = Permeability; Sh-Sw = Shrink-swell potential; Text = Texture.

Fig.1. High relief and contrasting landscapes contribute to the complexity of soil patterns in Waterton Lakes National Park and to the cartographic problems in report preparation.

true in the present case. Thus, a separation between Orthic Regosols and Cumulic Regosols on a given landform was seldom possible because drawing a line between the two would involve many field checks in areas where ground location on the photograph was difficult to ascertain. In general the mapped areas were named in terms of textural, stoniness and slope phases of their subgroup classification (Canada Soil Survey Committee, 1970). The legend was designed so that soil map area boundaries corresponded with landform boundaries such as alluvial fans, moraines, outwash plains, etc. Drainage, stoniness and other variations required that these larger landform units being subdivided.

The main objective of this survey was to provide interpretative information related to specific landscape locations for park managers and administrators in an easily comprehensible form. This dictated that once the soil information had been assimilated it could be easily interpreted for several proposed kinds of uses(e.g. picnic areas, trails, septic tank filter fields) and yet retain the desired amount of map detail (2–4 ha). There are two main kinds of legends that are commonly used to identify map units. The first, a multifactor legend, can be typified by the following set of symbols which were, at one time, suggested for use in Waterton Park:

$$\frown\ 1\text{–}3\ \mathrm{mc}\,\frac{2}{\mathrm{II}}\,T_3$$

Interpreted, this symbol indicates that the map unit is predominated by a number of knolls ($\frown$), rapidly to moderately well drained (1–3), medium coarse textured but gravelly and semi-compacted soils (mc $\frac{2}{\mathrm{II}}$), developed on glacial till represented by drumlins (T_3). This method is attractive because of its apparent ease of use. However, it was rejected because its open-ended nature allows for a phenomenal number of combinations of differing features which could not be easily interpreted. Although this symbolism provides a precise cataloging of soil features it does not provide a convenient presentation of data for subsequent interpretation because of the low number of repeating units. The second, a single-name, fixed-component kind of legend, is developed by setting up a symbol (name) with which to identify a central concept of the major soil and the extent of the variability associated with this central concept. The map unit descriptions provide the detailed information relating the soil characteristics and their variability to the map unit symbol. This second kind of legend was adopted, primarily because of the ease with which subsequent interpretations can be made. The concepts used to develop the legend were essentially those set out by Simonson (1963).

Because of the inability of the human mind to comprehend more than about 18 map units at one time (Simonson, 1971) map legends should be so designed that this number is not exceeded. If it can be assumed that the local park officials will want to examine relatively small areas in detail this objective has been fairly well met even though the legend is, in fact, about three times the maximum proposed. A generalization of the legend and the map will be necessary for effective use by the regional park office, where an overview approach is desired.

Mechanics of the field work

After a cursory examination to ascertain the main kinds of soils in the park a tentative legend was established based on the principles outlined in the foregoing discussion. Major landforms and associated soils were outlined on aerial photographs by stereoscopic interpretation and the delineated areas were named in terms of the tentative legend. By field checking it was possible to refine or subdivide the preliminary separations and to modify the legend to reflect the more detailed findings. An example of the kinds of separations is shown in Fig.1.

To allow for adequate characterization of the soils and to assist in the preparation of interpretive maps considerable time was spent sampling and describing representative pedons once the map units were firmly established. Supportive information such as infiltration and percolation rates, bulk densities and erosion characteristics of many of the map units were evaluated. The response of all the mapping units over which trails, roads or campgrounds have been built was recorded to assist in establishing guidelines for predicting the potential erosion hazard and other soil responses to imposed development. In areas such as Waterton, where few pertinent research data are available, the time and cost to adequately provide interpretive information and guidelines is considerably greater than in more studied areas. Some of the research findings and experience from Waterton will be transferrable to other mountainous national parks.

PRESENTATION OF THE INFORMATION

The ultimate worth of soil surveys in areas such as Waterton must be judged on whether they provide information which *saves* either needless expenditures or important natural resources. To *save* expenditures or resources, the inventory information must be presented in a simple, concise format which is amenable to use by park administrators. For this study the soil survey material will probably be used on one hand for relatively broad-scale planning and on the other for local management and on-the-spot evaluations. The following discussion will deal with the two situations separately.

Format of map presentation

For broad-scale planning the planners preferred that the soil survey information be presented on a planimetric base map to facilitate the use of transparent overlays (McHarg, 1969). They also preferred to have interpretations presented as generalized interpretive maps such as a map with degrees of susceptibility to erosion. Whether the loss of cartographic detail such as many of the wet areas (which are often 2–4 ha) will be of much concern in these interpretive maps has yet to be ascertained.

For local management purposes it was felt that soil survey information was best presented on a photographic base map (Fig.2). Available planimetric base maps provide insufficient cultural or physiographic information to locate many of the small map areas (Fig.3) precisely. However, most users can accurately locate, on the ground, areas delineated on photographic base maps. Thus, even without the ability to recognize the soils within the map units they are still able to avoid or to use the areas of special consequence for a given purpose.

The available semi-controlled photo mosaic for Waterton Park had mismatched areas large enough so that when transfer from the field photo was attempted some map areas were lost. Production of a mosaic of sufficient quality to avoid that problem proved to be impractical. Because the field information was recorded on about 100 alternate photographs it was undesirable to publish the information directly on them. The compromise adopted was to transfer the soil survey information to enlargements of small-scale photography so that 10 rather than 100 photo maps were needed. Photo centers were used but distortion was still too great to construct an acceptable mosaic. No attempt was made to make the lines on two adjacent photographs "join". They do, however, delineate the same

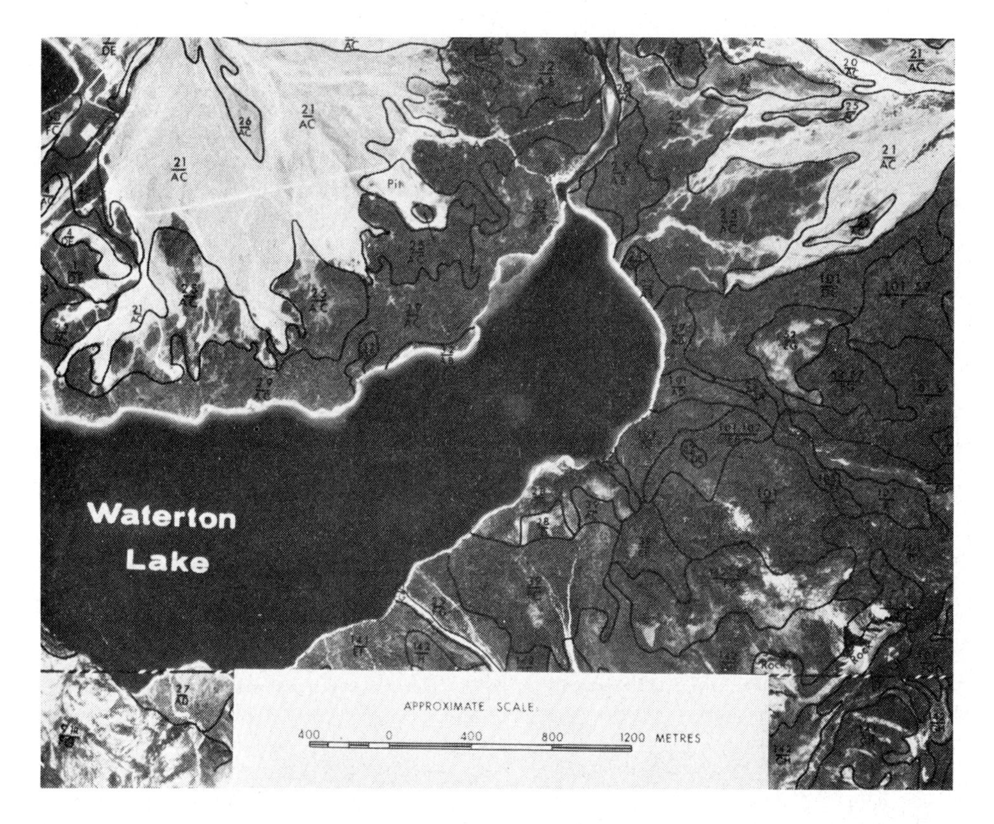

Fig.2. An excerpt from the soil map (Coen and Holland, 1973). The photographic base provides much information auxiliary to the soil data.

photo pattern and the same physical boundary on the landscape. Lines on adjoining photographs carry far enough into the joined area so that like areas (photo patterns) can be readily recognized. This procedure presented the information in a readily useable form but because of extreme distortion in places it violates the concept of quantified aerial presentation of information.

The proven usefulness of aerial photographic interpretation as a method of extrapolating soil boundaries (and boundaries in other natural resource inventories) ensures use of aerial photographs, almost universally, for recognizing and recording information on natural resources. The information must then be transferred from the field photographs to some final base for publication. This transfer posed special problems in Waterton Park.

Mechanical transfer from field photographs to another flight was found to be impossible in much of the park because of distortion. In many areas, the only way to transfer the boundaries was to follow like patterns on photographs from different flights, which is very time-consuming.

For effective comparison, information from two or more disciplines must be presented on the same base maps. Transfer of boundaries from field photographs to a planimetric

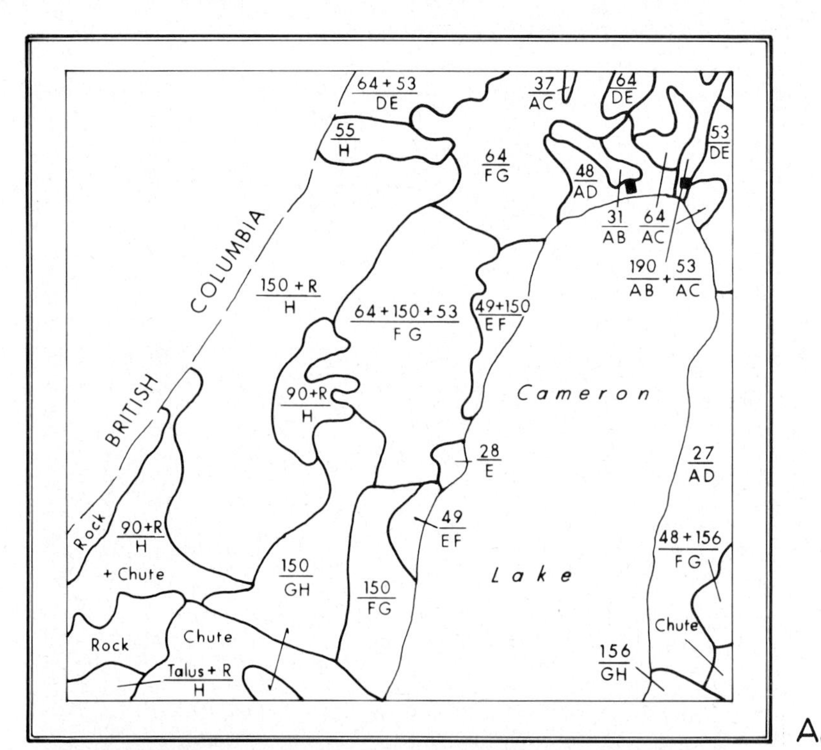

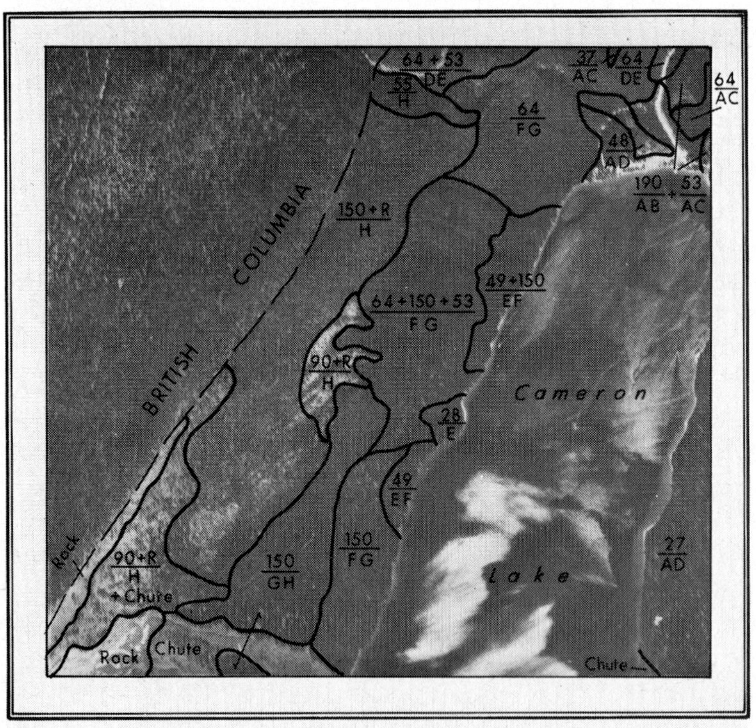

B

APPROXIMATE SCALE:

400 0 400 800 METRES

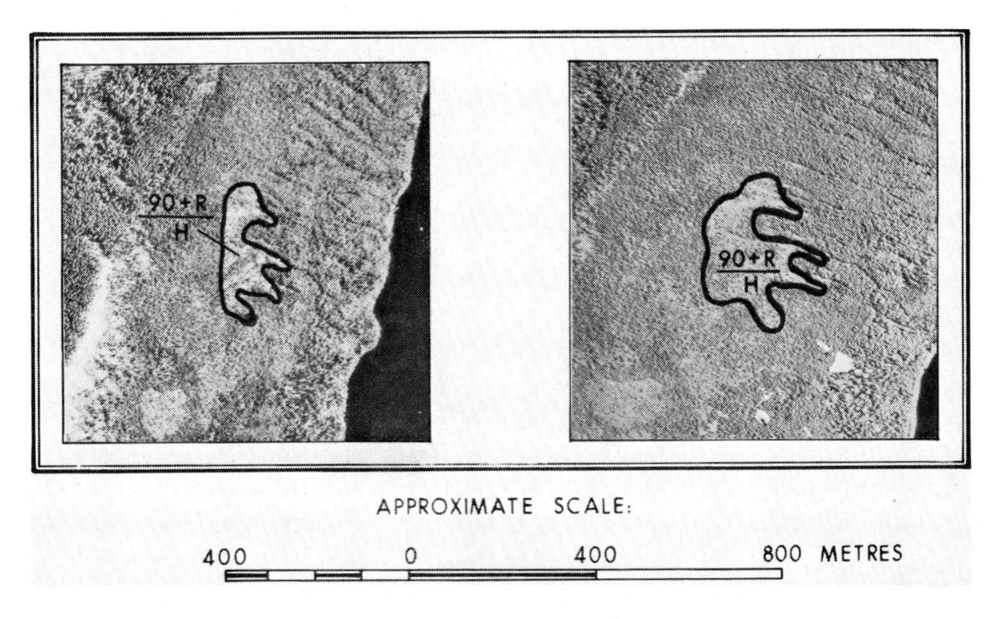

Fig.4. This shows the apparent change in the shape of the *90+R/H* area in Fig.3 when delineated on different photographs from the same flight. Unless distortion is corrected in some way, mechanical transfer from a photographic to a planimetric base can obscure correlations between related resources, e.g. soils and vegetation.

base was therefore attempted in Waterton Park. Small areas seemed to change considerably in size and shape, depending upon which photograph of a stereo pair had been used to record the boundary lines (Fig.3,4). Difficulties of transferring were especially great if forest cover and soils had been mapped on different alternate photographs. A given delineated area in which forest cover and soil boundaries coincided might not match at all after transfer to the planimetric base. This difficulty could be overcome, in part, by using stereoscopic plotting techniques for transferring both forest cover and soil boundaries. Stereoscopic plotting techniques permit averaging the distortion of a pair of photographs with overlapping coverage, allowing more accurate transfers of boundary lines. Experience in mapping of the park suggests that comparison and correlation of resource information from two or more disciplines is facilitated by using the same flight of photographs in the field work and the same techniques in transfers to a planimetric base. Even so, the mechanics of presenting information from detailed surveys of mountainous areas pose serious problems, especially if geodetic controls are limited.

Fig.3. Planimetric base (A) and photographic base (B) presentations of soil survey information. The area labelled *90+R/H* near centers of the figures is difficult to locate on the ground with the planimetric base because of the lack of cultural and physiographic features. Moreover, a 60% slope interferes with measurement of distance on the ground. The photographic base provides information on vegetation and landscapes which makes location of a desired area on the ground easier.

Format of the report presentation

A few guidelines for the nature and format of a report were available as a result of private communications with Natural and Historic Parks personnel but the usefulness of the adopted method of presentation will have to be assessed by the users of the report. The current impression is that the regional office for Natural and Historic Parks would like very detailed presentation of the written information (although generalized maps) with little technical terminology. On the other hand the local office would probably prefer to have mainly the interpretative information with little discussion and theory of obtaining the data. Large quantities of analytical data will be of little value to the local park personnel. In the report soon to be published (Coen and Holland, 1973) an attempt has been made to present the information so as to satisfy the needs of both user groups. Extensive use has been made of tables (Table I) and keys (Table II) with a minimum of essential text. The intent is that the report will act as a reference work permitting unneeded sections to be skipped without loss in continuity. An extensive table of contents has been included to facilitate this end. This allows description of the method of data acquisition and interpretation necessary to help avoid obsolescence without interfering with the ease of use of the report. Tables of analytical data are included in the appendices for the in-depth reader.

As mutual cooperation and understanding become better established between the user and producer of soil information the format of presentation will undoubtedly change. The nature and direction of the change will depend to a large extent on the patience and perception of the parties involved.

Interpretations of the soil information for recreation

Survey procedures, particularly the design of the legend, govern to a considerable extent the kinds of interpretations and ease with which they can be made. When multi-factor legends are used interpretive maps redrawn for each new interpretation are the only way to obtain the necessary information. In this method decisions necessary for generalizing the information are often made in the office by persons unfamiliar with the area mapped. These decisions are made more realistically in the field as the mapping proceeds. Thus, in fairly intensive surveys where specific interpretations are anticipated a single-name, fixed-component legend is much more efficient. This legend can have interpretations attached to the mapping symbol allowing predictions without any loss in cartographic detail. Interpretative maps can easily be drawn using interpretative tables such as Table I. Providing that detailed guidelines are established for the specific interpretations, it is possible to review the validity of the interpretations as the information is used, and by revising the guidelines, to update the interpretations as this is necessary.

A conscious effort has been made, throughout the project, to avoid suggesting that because the *soil* is suitable for a given use the *land* should be used for that purpose. Thus, evaluations were made in terms of *limitations for* a proposed use, or *susceptibility* to a given process affecting the use (Table I). This avoids implied suggestions, intended or otherwise, that land should be used for any given purpose.

TABLE II

Example showing the kind of information in the "Key to Major Characteristics of the Soil Mapping Units" used in Waterton Lakes National Park (Coen and Holland, 1973)

Soil Map Unit Number	Subgroup Classification[1]	Landform(s)	Parent Material	Main Horizons[1]	Texture and Coarse Fragments[1] (CF by volume estimate)
1	Orthic Dark Brown and Orthic Black Chernozems	Outwash plains, eskers, kames	Gravelly outwash	Ah, Bm, C	GSL, ~ 50% CF
15	Orthic and Cumulic Regosols	Floodplain	Sandy alluvium	Ah, C	SiL, <5% CF
17	Orthic Dark Brown Chernozem	Alluvial terrace	Gravelly alluvium	Ah, Bm, Cca	GSL, ~50% CF
21	Orthic Regosol	Alluvial fan	Gravelly alluvium	Ah, C	VGLS, >60% CF
31	Orthic Humic Gleysol	Alluvial fan	Loamy alluvium	Ah, Bg, Cg	Stratified SiCL-L, <5% CF
50	Orthic Dark Brown and Black Chernozems	Drumlins	Gravelly till	Ah, Bm, C	GL, ~ 40% CF
100	Cumulic and Orthic Regosols	Lower valley sides	Weathered fine shale	(Ah), C	SiC, < 5% CF
142	Orthic Regosol	Upper mountain sides	Coarse textured colluvium	Ah, C	GSL, GLS, 20-70% CF mainly fine gravels

Soil Map Unit Number	Internal Soil Drainage[1]	Main Topographic Classes[1]	Main Vegetation	Other Features
1	Rapidly drained	Complex slopes varying rapidly from A to G	Fescue, oat grass	Mainly esker area
15	Moderately well drained	AB	Willow, balsam poplar	Slightly elevated areas on river floodplains
17	Rapidly drained	AC	Fescue, oat grass	Weak Bm development
21	Rapidly drained	AC (occasionally steeper)	Fescue, oat grass	Braided channels
31	Very poorly drained	AB	Sedges	Saturated except for brief periods in the fall
50	Well drained	Short, complex slopes varying from A to G	Fescue, oat grass	Occasionally moraines
100	Well drained	Mainly simple slopes varying from A through H	Blue and brome grasses, timothy or saskatoon, fir and aspen (shrubby)	Grayish clays prone to slumping
142	Well drained	Mainly G and H slopes	Fescue, oat grass	>1500 m

[1]Descriptive terminology and classification according to Canada Soil Survey Committee (1970).

CONCLUSIONS

(1) Soil resource information for recreation areas should be based on properties of soil bodies.

(2) The information can subsequently be used to develop many interpretive evaluations.

(3) The kind of legend and the restrictions it imposes on the design (methodology) of the survey influences the ease with which use interpretations are made from the basic data.

(4) Klingebiel (1966) suggests that in areas of low use intensity 1 dollar spent on soil survey programs will return about 46 dollars over a 20-year period. The estimated cost of the Waterton Park survey amounts to about 4 cents per acre per year (1.6 cents per ha. per year) over a 25-year predicted life. Because of the intangible nature of many of the benefits it is difficult to predict a cost-benefit figure, especially at this early date. However, if a poorly located sewage lagoon has to be relocated because it is polluting nearby water this cost alone could justify the survey expenditures.

ACKNOWLEDGMENTS

The cooperation, assistance, and constructive criticism of W.D. Holland of the Canadian Forestry Service while jointly undertaking the survey is recognized. Many of the ideas expressed are rightfully his, but the responsibility for any errors remains that of the author.

REFERENCES

Canada Soil Survey Committee, 1970. *The System of Soil Classification for Canada.* Canada Department of Agriculture, Queen's Printer, Ottawa, Ont., 249 pp.

Coen, G.M. and Holland, W.D., 1973. *Soils of Waterton Lakes National Park and Interpretations.* Northern Forest Research Center, Canadian Forestry Service Information Report NOR-X-65, in press.

Klingebiel, A.A., 1966. Cost and returns of soil surveys. *Soil Cons.,* 32: 3–6.

McHarg, I.L., 1969. *Design with Nature.* The Natural History Press, Garden City, New York, N.Y., 197 pp.

Simonson, R.W., 1963. Soil correlation and the new classification system. Soil Sci., 96: 23–30.

Simonson, R.W., 1971. Soil association maps and proposed nomenclature. Soil Sci. Soc. Am. Proc., 35: 959–965.

Smith, G.D., 1965. Soil classification. *Pedologie*, 4: 134 pp. (spec.iss.).

Geoderma, 10 (1973) 87–98

APPLICATION OF PEDOLOGICAL SOIL SURVEYS TO HIGHWAY ENGINEERING IN MICHIGAN

K.A. ALLEMEIER

Testing and Research Division, Michigan Department of State Highways, Lansing, Mich. (U.S.A.)

(Accepted for publication June 26, 1973)

ABSTRACT

Allemeier, K.A., 1973. Application of pedological soil surveys to highway engineering in Michigan. *Geoderma*, 10: 87–98.

This paper describes the application of pedological soil surveys to highway engineering in Michigan. It includes an explanation of the organization, education, and training of personnel who perform this work and discusses the responsibilities and types of work performed by Michigan's District Soils and Materials Engineers. The types of information collected in soil surveys and their relation to the planning, location, and design of highways are discussed. Examples of soil type descriptions, pedological soil strip maps, design charts, and a resistivity survey profile are also included.

INTRODUCTION

Soil mapping in Michigan for highway purposes is done by the Department's own soils and materials engineers, using the pedological method of soil classification supplemented by point investigations where deemed necessary. This system has been used for the past 47 years. Information regarding the application of each soil series for highway design and construction, as acquired from experience, is available in the Department's *Field Manual of Soil Engineering* (fifth edition). For the purposes of this paper, it is not deemed necessary to trace the history or explain the theory of the pedologic system of soil classification (Simonson, 1962). This system, with or without modifications, is now being used by many highway organizations in the United States.

PERSONNEL ORGANIZATION AND TRAINING

Michigan has nine highway districts, varying in size from 4 to 13 counties, with 687 to 2,182 kilometers of State trunkline highways. In each district there is one district soils and materials engineer with one or two assistant soils engineers or soils technicians, depending upon the work load. These men are responsible for all of the soils engineering work in their particular district. In general, this consists of: reconnaissance soil appraisals for route location; soil surveys; peat and rock soundings for design purposes; location and investigation of borrow sources (not including sources of produced aggregates); and methods of waste disposal for rest areas. During the construction stage, it is also the soils and materials engineer's responsibility to investigate the following: subgrade for potential frost heave and differential heave areas, seepage zones, and water tables; quality of sand subbase; quality of compaction; and the methods of treatment of peat swamps. Beyond that, he is expected to recognize and assist in the solution of embankment and structure foundation problems and to provide soils information which may be necessary for maintaining the highway system.

The central soils office in Lansing correlates this information; acts as liaison between the Divisions of Planning, Design, Construction, and Maintenance; establishes policies; and offers advice and administrative assistance. Scheduling and performing the routine work are the responsibility of the individual soils and materials engineer. The special point or area investigations, which are not performed by the district soils and materials engineer, are foundation borings and resistivity and seismic surveys. The foundation boring work is the responsibility of the central soils office, whereas resistivity and seismic surveys are carried out by the central Testing Laboratory.

There are 25 engineers in the Soils and Materials Section, including men in the central office and at the district level; 21 have degrees in civil engineering, and 4 have degrees in geology. Of the 25 engineers, 23 are registered professional engineers. In addition to these engineers, there are three technicians who assist in the district work. These men are former foundation boring crew chiefs.

In general, the policy is to obtain men with civil engineering degrees and train them for the soil mapping work. Usually, they are assigned to a district which will have one or two experienced soils and materials engineers. They assist in mapping four or five projects along with the district soils and materials engineer, and from the experience gained in this manner can start making soil surveys on their own. In case they run into soils with which they are not familiar, they can call upon the district soils and materials engineer for assistance in mapping these areas. An invaluable aid in gaining soil mapping experience is to map a project and then, during the construction stage, to study the exposed profiles and correlate them with their original maps.

Each summer, in selected areas of the State, soil classification tours are conducted by the Soil Science Department of Michigan State University; primarily for local county extension agents, agriculture teachers, and the local U.S. Department of Agriculture (USDA) soils mapping personnel. These tours are also attended by the highway soils and materials engineers and aid in their training and knowledge of pedological classification. If geological tours are held, they are usually attended by the local soils and materials engineers to aid them in identification of landforms and associated soils.

PREPARATION OF A PEDOLOGICAL SOIL STRIP MAP

To illustrate the soil mapping procedure as used in Michigan, the various stages through which a project is followed by the soils and materials engineer will be discussed. The first use of soil mapping is by the Planning Division in its long-range program studies. The soils and materials engineer is consulted as to whether there may be any soil factors influencing the cost of the job and affecting long-range budgeting. At this stage, these factors consist primarily of long or deep swamp deposits which cannot be avoided, bedrock, extensive areas of soft lacustrine clays requiring special treatment, and longer than average haul distances for borrow materials.

The next stage in which soil information is used is during the route location study. During this study, the Route Location Section makes use of the existing USDA county soil maps and the State geological maps in trying to avoid swamp sections and to utilize borrow sources. Where necessary, these are often supplemented by more detailed soil classification or peat soundings obtained in the field by the soils and materials engineer. This study for route location may cover an extensive area and, therefore, is not made in the detail which is used after the final line has been selected. Soil information is also an important part of that on environmental factors which may be affected by highway location.

After the final location has been established and the line staked by the survey crew, the complete detailed soil survey is made. It is made at this time so that the Design Division will have full use of all necessary soil information in selecting the preliminary design and grade. The special point or area investigations will be made at a somewhat later date.

The base which is used for the soil map will vary, depending on what is available for the project. For projects on which aerial surveys have been made, aerial mosaics will be available at a scale of 2.54 cm to 61 m and these are often utilized as base maps. The photographs for these mosaics are obtained by special line flights. The aerial base map is a positive print on photographic paper. Advantages of the aerial photos are: (1) topographic features are readily determined and the soil survey can easily be made even if the survey stakes have been removed; (2) soil boundaries are easier to define; and (3) drainage characteristics are more apparent. The main disadvantages are the small scale of the photo and they are somewhat bulky to use in the field. Furthermore, the scale is smaller than that used by the Design Division in the preparation of plans and the information cannot be traced directly. The soil boundaries as shown on the aerial photo are transposed to the plan sheet, using the stationing shown on the aerial photo.

Where aerial photographs are not available, the most common base sheets are 21.6 by 27.9 cm cross-section paper at a scale of 2.5 cm to 30.5 m. The plan sheet may be used as a base map if the survey has already been plotted.

Once the base map has been selected, the soils and materials engineer is ready to prepare the soil map. Before beginning the detailed field survey, several sources of information are studied. One of these is the State surface geology map. With this, he can readily determine which geologic landforms and associated soils may be encountered. If the USDA county soil map is available, this is also studied. The older soil maps do not present enough detail for highway mapping purposes, but they do assist in narrowing down the number of anticipated soil series. Also available are small-scale aerial mosaics of the entire State, although at present Michigan does not utilize them to any great extent in interpreting soil surveys. They may be studied, however, as an indication of what to expect. The major use of these available aerial mosaics by the soils and materials engineer

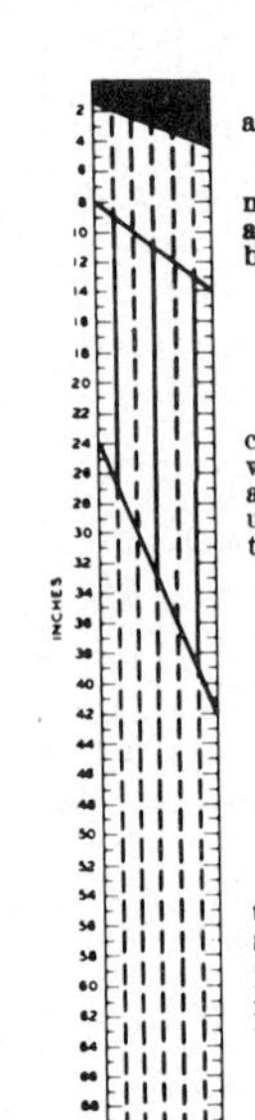

Litter, leaf mold and humus.

Pale brown loam, mottled with gray and yellowish brown.

Yellowish brown clay loam, mottled with gray, yellow, and orange, subangular blocky structure.

Calcareous brown to light brownish gray loam or silt loam with light gray yellowish brown and yellow mottling.

CONOVER

SERIES DESCRIPTION

The surface texture is a loam, a fine sandy loam or a silt loam.

The imperfectly drained Conover soils developed on loam or silt loam till materials. They occur on nearly level undulating lake and till plains in southern and central Michigan. The substratum texture is intermediate between the Blount (clay loams) and the Locke (sandy loams) soils.

Conover is the imperfectly drained member of the catena which includes the well-drained Miami, the moderately well-drained Celina and the poorly drained Brookston soils.

Crosby soils have similar drainage and textural profile, however, the surface layer has less organic matter content and consequently has a light gray color. In mapping, include Crosby with Conover soils.

Celina soils where they occur on till plains have mottling at 18 to 36 inches. In many detailed surveys Celina has been combined with Conover as a complex. When mapping on till plains, include Celina with Conover soils.

Metamora and Macomb soils comprise 20 to 40 inches of sandy loam materials which may have a loam subsoil over typical Conover clay loam subsoil and loam parent materials. The upper part of the Macomb soils are gravelly loam, water deposited materials. Therefore, include Metamora and Macomb soils with Conover in mapping.

Capac and London soils have similar textural and drainage profiles as the Conover soils. Capac and London soils occur on till and lake plains in the Thumb Area and in north central Michigan. In mapping, include Capac and London soils with the Conover soils.

A representative mechanical analysis of the substratum follows:

2.1%—Very coarse sand
2.6%—Coarse sand
4.6%—Medium sand
10.1%—Fine sand
8.5%—Very fine sand
47.9%—Silt
24.2%—Clay

Construction Information

Excavation in this material is not generally difficult. In wet periods the material will be slippery and hauling difficult. The surface will crust and become hard in periods of prolonged hot dry weather. Seepage may be encountered, but not extensive enough to be a serious construction problem.

CONOVER

Litter, leaf mold and humus.

Light yellowish brown loam or sandy loam.

Dark brown to dark yellowish brown clay loam, angular blocky structure.

Light yellowish brown calcareous loam till, massive to subangular blocky structure.

MIAMI

SERIES DESCRIPTION

The surface texture is usually a loam or a fine sandy loam.

The well-drained Miami soils developed on loam or silt loam till materials. They occur on nearly level to hilly till plains and moraines in southern and central Michigan. The substratum texture is intermediate between the Morley (clay loams) and the Hillsdale (sandy loams) soils, and may contain occasional layers and pockets of sand, silt and massive clay. Seepage frequently occurs in these pockets.

Miami soils are the well-drained members of the catena which includes the moderately well-drained Celina, the imperfectly drained Conover and the poorly drained Brookston soils.

Owosso, Kendallville and Cadmus soils are associated with Miami soils especially on moraines. Owosso soils comprise 20 to 40 inches of sandy loam materials over typical Miami subsoil and substratum textures. The upper part of Kendallville and Cadmus series are gravelly loam outwash materials. Cadmus soils are moderately well-drained. On many county surveys the mapping unit is designated as an Owosso-Miami complex. Therefore, include Owosso, Kendallville, and Cadmus soils with Miami soils in mapping.

Marlette and Guelph soils have similar drainage and textural ranges in the subsoil and substratum as the Miami soils. They have comparatively thin ashy gray and brown upper horizons which are characteristic of the soils in northern Michigan. The ashy gray horizon can be found only in areas that have not been cultivated. The calcareous substratum of the Guelph soils occurs at 15 to 25 inches. Marlette and Guelph soils occur on moraines, largely in the Thumb area. In mapping, include Marlette and Guelph soils with Miami soils.

A representative mechanical analysis of the substratum follows:

8.0%—Very coarse sand
4.4%—Coarse sand
7.2%—Medium sand
16.8%—Fine sand
10.8%—Very fine sand
29.8%—Silt
23.0%—Clay

Construction Information

Excavation is generally not difficult. In wet periods, the material will be slippery and difficult to haul over. The surface will crust over and become hard in periods of prolonged dry weather. Seepage may be encountered but usually is not extensive enough to be a serious construction problem.

MIAMI

Fig.1. Soil series profiles from *Field Manual of Soil Engineering* (5th ed.).

Litter, leaf mold and humus.

Very dark gray loam, fairly high in organic matter content.

Dark yellowish brown to yellowish brown clay loam, mottled with gray, pale brown and orange, usually blocky structure.

Yellowish brown calcareous loam, mottled with gray and pale brown. Massive structure in places.

BROOKSTON

SERIES DESCRIPTION

The surface texture is a loam or a silt loam.

The poorly drained Brookston soils developed on loam or silt loam till materials. They occur on level to nearly level till plains in central Michigan and on lake and till plains in the Saginaw Bay area and in eastern and southeastern Michigan. The substratum textures are intermediate between the poorly drained Pewamo (clay loams) and the poorly drained Barry soils (sandy loams).

Brookston soils are the poorly drained member of the catena which include the well-drained Miami, the moderately well-drained Celina, and the imperfectly drained Conover soils.

Corunna and Berville soils comprise 20 to 40 inches of sandy loam materials which may have some textural development over typical Brookston subsoil and substratum textures. The upper part of the Berville soils are gravelly loam outwash materials. Therefore, include Corunna and Berville soils with Brookston soils in mapping.

Parkhill and Tappan soils have similar textural and drainage properties as Brookston soils. These soils occur on till and lake plains in the Thumb Area and in north central Michigan. In mapping, include Parkhill and Tappan soils with Brookston soils.

Kokomo soils are similar in textural profile, and geologic origin to the Brookston soils, however, they developed under poorer drainage conditions which is reflected by a higher percentage of organic matter in the surface layer, which is deeper and darker-colored (mucky). In mapping include Kokomo soils with Brookston soils.

A representative mechanical analysis of the substratum follows:

1%—Very coarse sand
1%—Coarse sand
2%—Medium sand
10%—Fine sand
15%—Very fine sand
48%—Silt
23%—Clay

Construction Information

This material is poorly drained and is usually wet in its natural state. Excavation in this material is uncommon except for ditches, removal of top soil, and depressed sections of the highway. This material is soft and slippery when wet and hauling over these areas when in this condition is difficult.

BROOKSTON

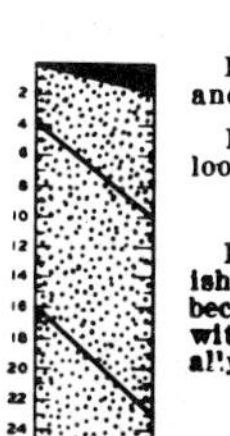

Litter, leaf mold and humus.

Pale to ashy gray loose sand.

Brown to yellowish brown loose sand becoming lighter with depth. Usually not cemented.

Grayish yellow, pale yellow or light brownish yellow loose sand with occasional layers, pockets and lenses of silts, very fine sands, clay loams, sandy clay loams and sandy loams. Some gravel cobbles and boulders.

ROSELAWN

SERIES DESCRIPTION

The surface texture is usually a sand.

Roselawn soils are well-drained, deep, sandy soils which developed on rolling to extremely hilly moraines largely in the northern part of the Lower Peninsula.

The original forest was dominantly pine. In places, stones and boulders are imbedded in the soil and scattered over the surface. Layers and pockets of silt, sandy clay and very fine sand may be found at any depth causing a perched water table. Normally the water table is deep.

Roselawn soils are similar to Wexford soils in textural profile, drainage, and geologic origin but are distinguished by having less limestone influence, and by the original pine vegetation. Wexford soils also have a darker colored (reddish brown) subsoil with more organic matter than the Roselawn soils.

Roselawn soils are similar to Rubicon and Grayling soils in textural profile and drainage, however, Roselawn soils occur on morainic areas, with pockets, layers, and lenses of finer-textured materials common at 4 to 20 feet. Rubicon and Grayling soils occur on outwash plains.

A representative mechanical analysis of the substratum follows:

4%—Very coarse sand
28%—Coarse sand
42%—Medium sand
22%—Fine sand
2%—Very fine sand
1%—Silt
1%—Clay

Construction Information

This material is considered excellent for grading operations during any season of the year. However, the loose character of the sandy material may interfere with hauling operations. Seepage may be encountered in the sandy clay and silt pockets in sufficient amounts to make construction difficult in local areas.

ROSELAWN

is during the stage of borrow study and location. However, most borrow materials are now located and furnished by contractors.

After this available information has been studied, the soils and materials engineer is ready to begin the detailed soil survey. Tools used for this purpose consist of a tiling spade, orchard or bucket auger, and a worm auger. As he progresses with the soil survey, the engineer constantly observes the topography, the landform, the vegetation, and the agricultural use of the land. Each of these provides valuable tips in soil mapping. He then proceeds to examine the soil profile. This is done by exposing the profile in a cut section or side hill or by boring a hole. As the material is removed, he lays it out in the form of a soil profile and he notes the texture, color, consistency, and the depths at which the various horizons are encountered. From the information which has already been gained by studying the county soil map and the geologic landforms, he can now anticipate perhaps five or six different soil series. Then, by using the soil profiles in the Department's *Field Manual of Soil Engineering,* and studying the exposed profile, he can identify the soil series that he is examining (Fig. 1).

As the pedological system shows area significance, it is not necessary to bore holes at certain specified intervals or to specified depths. Conditions at the site determine the spacing and depth of the borings. Only enough profiles are exposed to identify the soil series being mapped and determine its boundaries. This is based on the judgment and experience of the soils and materials engineer. Soil boundaries are quite often recognized

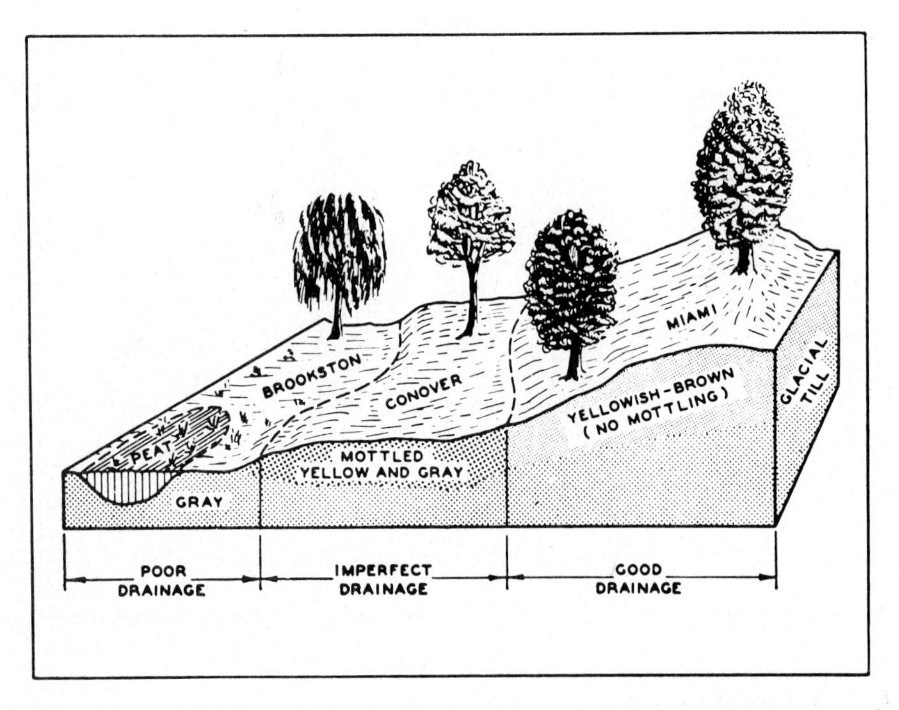

Fig. 2. The drainage and relief characteristics of a soil catena (Miami–Conover–Brookston shown).

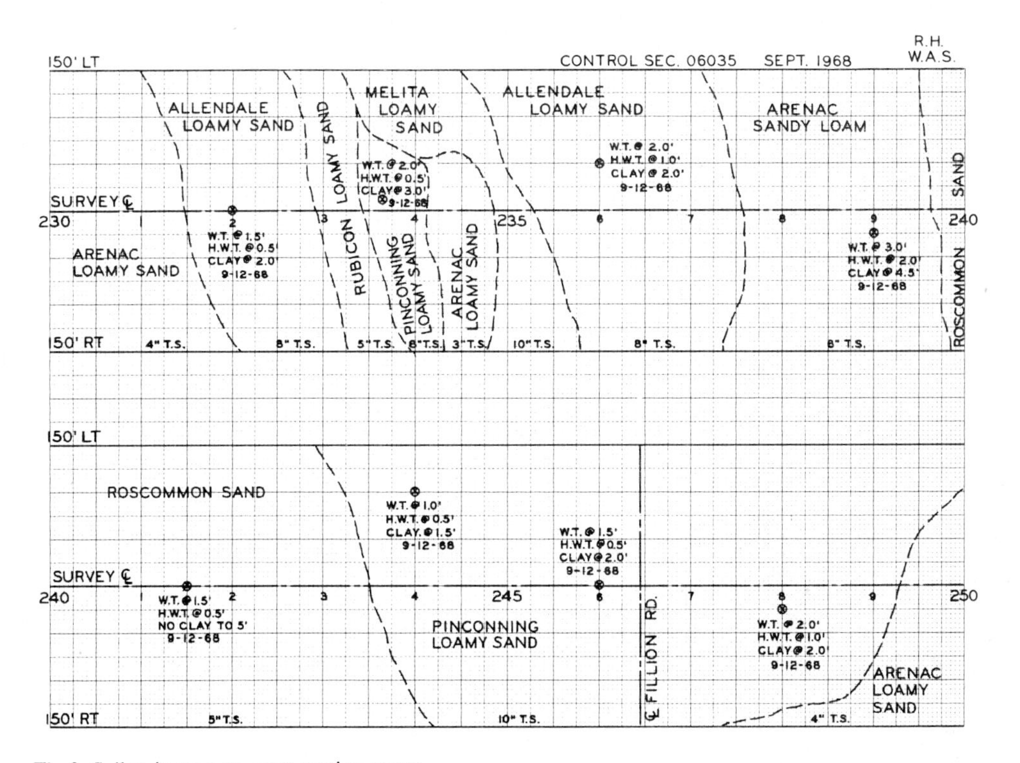

Fig. 3. Soil strip map on cross-section paper.

from changes in topography, changes in vegetation, or changes in the color or feel of the natural ground (Fig. 2). The entire area within the proposed right-of-way is mapped.

At each boring, the soils and materials engineer will record the depth of topsoil. However, when the final soil map is submitted, the topsoil depths may be submitted either for each boring or as an average depth within the boundaries for an entire soil series. Depending on the soil series mapped, the soils and materials engineer may obtain the depth to water table, underlying clay, or bedrock. When water tables are mapped, he not only shows the existing water table, but also the expected high water table which can be determined by the color and the amount of mottling in the soil. Water tables are observed at intervals of 61–122 m. As the field soil survey progresses and boundaries are mapped, the soils and materials engineer notes sites where point investigations will be required. Mapping in this manner, the experienced soils and materials engineer can map from two to five miles per day.

On the map which is being made, the soils and materials engineer will show the boundaries between soil series by dashed lines and denote the name of each soil series. Also on this map he shows depth of the topsoil and the depth to water table, underlying clay, and bedrock, if encountered (Fig. 3).

DESIGN CHART — SOIL ENGINEERING DATA AND RECOMMENDATIONS

Soil Series	Characteristics: Brief Description of Typical Profile (See Appendix E of Field Manual of Soil Engineering for Complete Description)	Characteristics: Adapted to Winter Grading	Characteristics: Normal Depth to Water Table	Treatment – Grade: Recommended Location of Plan Grade with Respect to Natural Ground	Treatment – Grade: Estimated Percent of Boulders (Rock Excavation)	Treatment – Grade: Estimated Depth of Topsoil (ft)	Treatment – Grade: Subbase Recommended (See note a)	Treatment – Grade: Estimated Lineal Feet Per 1000 Feet of Cut Below Natural Ground Elevation (b): Subgrade Undercutting (Per Roadbed)	Treatment – Grade: Estimated Lineal Feet Per 1000 Feet of Cut Below Natural Ground Elevation (b): Edge Drains (Per Roadbed)	Treatment – Grade: Bank Drains: Use only if cut is deeper than: / Lineal Feet	Treatment – Grade: Granular Blanket: For Cuts Deeper Than 6.0 Ft. Percent of Slope Area	Treatment – Embankment: Suitable Borrow for Embankment Construction	Treatment – Embankment: Percent of Shrinkage	Resources: Granular Material Class II	Resources: Possible Source of Gravel	Resources: Source of Topsoil	Remarks
		1	2	3	4	5	6	7	8	9	10	11	12	13	14	15	
Brookston	Poorly drained loam or silt loam	No	Shallow(d)	Fill 4'-5'(g)	0.0	0.7-1.0	Yes	300(k)	600(k)		5(k)	No	25-35	No	No	Excellent	
Conover	Imperfectly drained loam or silt loam	No	Indefinite(e)	Remarks	0.0	0.6-0.9	Yes	300	500		5	Ltd.	25-35	No	No	Excellent	(3) Controlled by surface drainage.
Miami	Well drained loam or silt loam	No	Deep	Anywhere	0.1	0.4-0.7	Yes	300	600		5	Yes	25-35	No	No	Good	
Roselawn	Well drained sand	Excellent		Anywhere	0.0	0.1-0.4	No(m)	200	300		2	Yes	10-20	Yes	No	No	

General Notes

(a) Applies when the grade lies in cut or when the embankment is constructed of material from the same series. Subject to compliance with specification gradation and extent of area within project.
(b) Lineal feet of cut to be measured along a line below plan grade determined by the maximum depth of subgrade undercutting specified.
(c) Fine grading difficult due to stoniness.
(d) Shallow: Indicating that at times this series will be under water, but after the water has receded there is no apparent water table.
(e) Indefinite: No true watertable. Possible seepage water at any depth.
(f) Extremely variable due to its origin.
(g) The higher grade heights are a minimum standard for primary trunklines, expressways and interstate routes.
(h) A fill will always be required due to the wet condition of the soil and the possibility of overflow.
(k) Applies where standards of vertical alignment require cut sections, in variance with recommendations in column 3.
(m) Subbase recommended if grade line is in "B" horizon (upper 3' of soil profile).
(n) Additional excavation required for transition from cut to fill and in shallow cut and fill areas, as per sketch in Field Manual of Soil Engineering, Fig. 3.31.

Fig. 4. Excerpt from the Design Chart, *Field Manual of Soil Engineering* (5th ed.).

DESIGN RECOMMENDATIONS

The soil strip map, along with recommendations pertinent to the project, are transmitted to the Engineer of Soils and Materials in the central office. Recommendations for design purposes are usually not made in detail. These are provided in the Department's *Field Manual of Soil Engineering,* particularly its soils Design Chart which has recommendations concerning the engineering properties for highway design of each soil series. The Design Chart is the result of intensive correlation of experience and performance of each soil series since the pedological system was adopted in Michigan. It is revised periodically to conform with new policies and specifications. The recommendations can be applied for each soil series no matter where it is mapped in the State, and the only recommendations which need be submitted with the soil strip map are those which vary from the recommendations given in the Design Chart (Fig.4). Pavement type and cross-section thickness are extremely dependent upon the supporting capacity of the soil, and the soil series and texture are important factors in pavement design.

The Department does not recognize all soil series mapped by the USDA and the Michigan Agricultural Experiment Station in the State. Many soils which have only slight differences in texture, drainage, pH values, etc., are grouped together under one series. However, the 165 series mapped by the Highway Department have the same names and essentially the same profile descriptions as those adopted by the USDA. Recognition and combination of series are correlated with the Soil Conservation Service. Although 165 series may seem like a large number with which to be familiar, it must be recognized that many series are found in only certain parts of the State and that with the information available, and the soils and materials engineer's experience, the anticipated number of soil series on a particular project can be narrowed down to 10–12.

POINT INVESTIGATIONS

As previously mentioned, the district soils and materials engineer ordinarily does not make extensive point investigations during the preparation of the detailed soil strip map. However, while the project is still in the preliminary design stage, he will return for these investigations.

The type of investigation which occupies the majority of his time is peat sounding. He refers to the soil survey for these areas and boundaries. Soundings are taken at 15.3–30.5 m intervals along the line and at not less than 15.3 m intervals across the grading section. The soundings are usually made by hand with a Davis Peat Sampler. As the soundings are taken, the depths at which changes in the peat classification occur are recorded. Experience has shown that in many swamps, sand layers may be encountered which cannot be penetrated by hand tools, thus giving an impression of a true bottom. To prevent this, in swamps more than 152.5 m in width,supplemental borings are made at intervals of 61–91.5 m to determine whether the swamp may have a false bottom. This is done by two special crews which operate out of the central office, using lightweight hydraulic boring equipment.

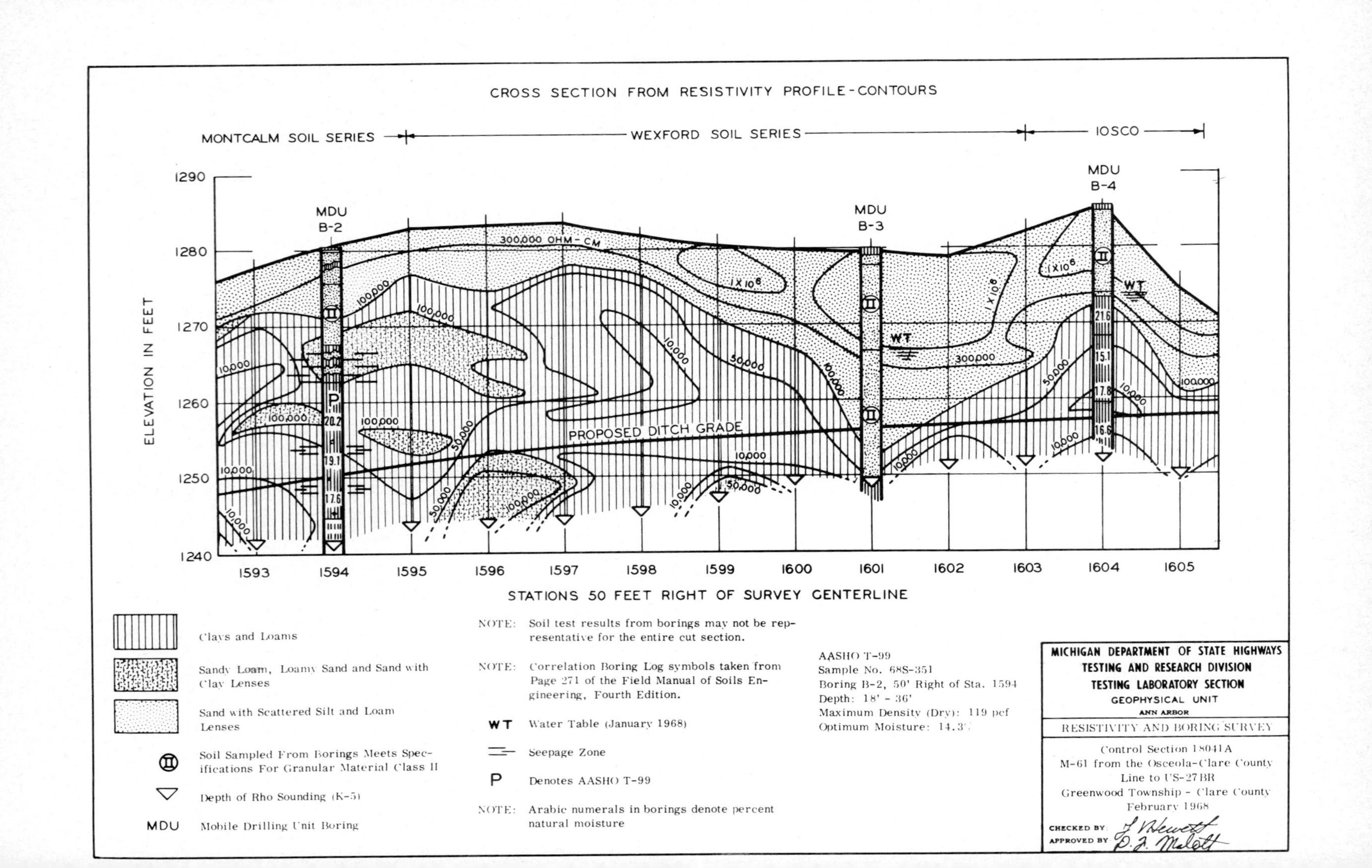

CROSS SECTION FROM RESISTIVITY PROFILE-CONTOURS
MONTCALM SOIL SERIES
WEXFORD SOIL SERIES
IOSCO
ELEVATION IN FEET
1290
1280
1270
1260
1250
1240
MDU B-2
MDU B-3
MDU B-4
300,000 OHM-CM
PROPOSED DITCH GRADE
WT
1593
1594
1595
1596
1597
1598
1599
1600
1601
1602
1603
1604
1605
STATIONS 50 FEET RIGHT OF SURVEY CENTERLINE
Clays and Loams
Sandy Loam, Loamy Sand and Sand with Clay Lenses
Sand with Scattered Silt and Loam Lenses
Soil Sampled From Borings Meets Specifications For Granular Material Class II
Depth of Rho Sounding (K-5)
MDU Mobile Drilling Unit Boring
NOTE: Soil test results from borings may not be representative for the entire cut section.
NOTE: Correlation Boring Log symbols taken from Page 271 of the Field Manual of Soils Engineering, Fourth Edition.
WT Water Table (January 1968)
Seepage Zone
P Denotes AASHO T-99
NOTE: Arabic numerals in borings denote percent natural moisture
AASHO T-99
Sample No. 68S-351
Boring B-2, 50' Right of Sta. 1594
Depth: 18' - 36'
Maximum Density (Dry): 119 pcf
Optimum Moisture: 14.3%
MICHIGAN DEPARTMENT OF STATE HIGHWAYS
TESTING AND RESEARCH DIVISION
TESTING LABORATORY SECTION
GEOPHYSICAL UNIT
ANN ARBOR
RESISTIVITY AND BORING SURVEY
Control Section 18041A
M-61 from the Osceola-Clare County Line to US-27BR
Greenwood Township - Clare County
February 1968
CHECKED BY:
APPROVED BY:

Rock soundings are also the responsibility of the district soils and materials engineer. Where possible, these are made with a truck-mounted power auger. At every station, at centerline, and at the ditch lines these borings are made to a depth of 1.2 m below the proposed plan and ditch grades. In areas inaccessible by truck, a Barco Hammer, driving 2.54 cm steel rods is used. Rock soundings are often made with seismic or resistivity methods.

Resistivity and seismic surveys are used extensively in Michigan for area investigation. In all cuts more than 3–3.7 m deep, resistivity surveys are made. Seismic apparatus is used primarily where it is anticipated that bedrock may be encountered. These surveys are made by a special section which operates from the central Testing Laboratory. Wherever correlation borings are made, samples are submitted for mechanical analysis and determination of Atterberg limits. This information is shown on the resistivity report (Fig.5).

Foundation borings for larger culverts, bridges, and embankments are not the responsibility of the district soils and materials engineer. Foundation borings are made by special boring crews from the central office.

CONCLUSIONS

In Michigan, one of the advantages of the pedological system is that it has been developed over a long period of years and is backed by extensive field correlation, experience, and performance. The qualities peculiar to each soil series are summed up in its name and are easily recognized by design and construction engineers and contractors. Moreover, mention of a soil series name brings to mind the geology, topography, texture, and drainage characteristics of the particular soil. Thus, a poorly drained soil can be recognized at any time of the year. The soil series name also expresses a three-dimensional concept in that it not only describes depth but also area significance along and on either side of a survey line. Since one man can map from two to five miles per day, the system can be applied not only to major highways but to all reconstruction projects in a highway program. One of the major differences between the pedological system and other methods of soil classification is that in the pedological system, laboratory investigation and testing supplement field investigations. In other systems the field investigations furnish the information for the laboratory classification.

The reliability of the information depends to some extent upon the experience and ability of the soils and materials engineer who prepares the soil strip map and upon the amount of detail shown. As has been mentioned, the Michigan Design Chart is a result of years of correlation and experience. Errors which might result from incorrect interpretations of the Design Chart are usually prevented by a close liaison between the central soils and materials office and the Design and Construction Divisions

Fig.5. Typical resistivity survey.

REFERENCES

Michigan Department of State Highways, 1970. *Field Manual of Soil Engineering.* Lansing, Mich., 474 pp.
Simonson, R.W., 1962. Soil classification in the United States. *Science,* 137: 1027–1034.

Geoderma, 10 (1973) 99–112

TWENTY-FIVE YEARS OF APPLICATION OF SOIL SURVEY PRINCIPLES IN THE PRACTICE OF FOUNDATION ENGINEERING

G.D. AITCHISON

Division of Applied Geomechanics, Commonwealth Scientific and Industrial Research Organization, Syndal, Vic. (Australia)

(Accepted for publication October 19, 1973)

ABSTRACT

Aitchison, G.D., 1973. Twenty-five years of application of soil survey principles in the practice of foundation engineering. *Geoderma*, 10: 99–112.

The evidence of many years suggests that the principle of a coupled pedological soil type and foundation engineering practice is useful and may continue to be applied. Considerations of economy are well met by this approach which has already permitted at least 10,000 foundation assessments following the detailed study of only 33 soil types. However, recent knowledge expressed in terms of a more sophisticated type of soil mechanics behaviour suggests the wisdom of separate quantitative assessments of soil response in all circumstances in which an "abnormal" soil moisture regime might be anticipated.

INTRODUCTION

Domestic and institutional buildings in the State of South Australia have been constructed, traditionally, with walls of brick or masonry. Such structures are of a stiff-brittle type and are particularly sensitive to distortions arising from differential movements of portions of the foundations.

On clay soils which may be exposed to moisture changes, such differential movements may reach significant proportions leading to unsightly and unacceptable cracking in the structures. The acceptable magnitude of deformation and/or cracking in the walls of domestic buildings is a matter of aesthetics – which will vary from one population or area to another – but in the circumstances of this study it was evident that the requirement of the majority of the population was for a totally crack-free structure.

Experience, and some relevant studies, had shown that such a crack-free structure – of the traditional type of construction – could only exist if the ratio of differential vertical movement to length of wall affected by such differential movement did not exceed 1/1,000.

Despite such an idealised requirement, it was apparent – at the commencement of this study in 1945 – that a very large proportion of the traditional type of domestic buildings erected on clay soils in the populated areas of South Australia exhibited a pattern of cracking of disconcerting proportions. (Many homes showed angular distortions approaching an order of magnitude greater than that corresponding to the crack-free state. Cracks in walls often exceeded 1 cm in width while in a small number of houses cracks greater than 5 cm in width were observed.) However on other soils, the same types of structures were entirely sound.

Since so many structures were seriously affected by this phenomenon a program of research was initiated in an attempt to define the specific causes of foundation failure, the types of soils contributing to such failure, the quantitative relationships (if any) between soil types and those soil properties affecting foundation performance, and finally the relationships (if any) between soil type and acceptable foundation practice.

The study area chosen was the urban and suburban region of the City of Adelaide (capital of South Australia) – a total of approximately 900 km^2. Adelaide is located at 35°S and 138.5°E. It is situated in an uplifted alluvial plain between the Mount Lofty Range and St. Vincent Gulf. The climatic environment of the Adelaide plains is that of a warm temperate zone – typically Mediterranean with a pronounced seasonal cycle of winter rainfall and summer drought. The monthly mean minimum temperature during winter is approximately 8°C, whereas the monthly mean maximum temperature during summer is approximately 27°C. Average annual rainfall is approximately 60 cm.

RESULTS OF STUDY

The detailed results of the study have been reported fully (Aitchison et al., 1954). In brief, the principal points of relevance which were established during the study included the following.

The dependence of foundation characteristics upon the total (pedological) soil profile

The cheapest and simplest foundation practice adopted (traditionally) within the study area was, and still is, the surface strip footing (see Fig.1). Such a footing, or equally a continuous concrete slab footing, on or slightly beneath the soil surface can be expected to reflect the whole of the physical response of the total soil profile to a change in environmental controls. Thus, if any layer tends to soften, due to wetting, and thus to yield under foundation loadings, the response of a building supported on a surface strip or slab should reflect the total of all such yields of all layers in the soil profile. Similarly, if any layer should exhibit shrinkage or swelling behaviour, the response of a structure supported on a surface footing should reflect the total of all such shrinkage or swelling movements in all layers of the soil profile.

Extensive observations of soil profiles throughout the study area and of foundation performances throughout the same area showed, in fact, that a high degree of correlation existed between the total soil profile and foundation experience as demonstrated by the presence (or absence) and the severity of cracking in typical buildings supported upon surface footings.

By contrast, an alternative type of footing – the bored pile or pier-and-beam (see Fig.1)

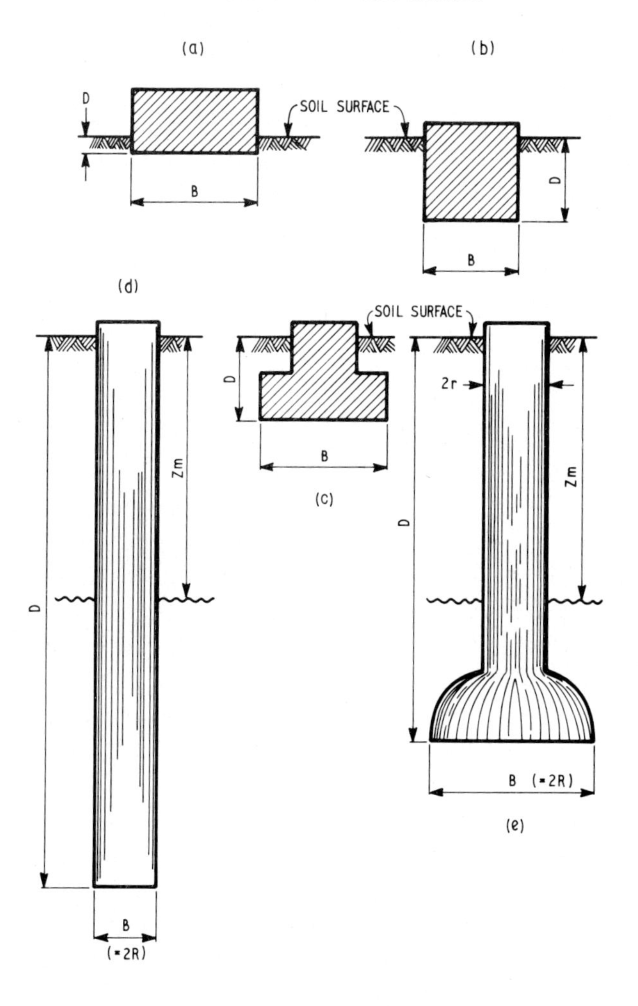

Fig.1. Foundation practices on the soils of Adelaide and suburbs.(a), (b) and (c): variants of "surface strip footings"; (d): bored pile; (e): under reamed pier.
Types (d) and (e) support beams suspended clear of the soil surface. Z_m represents the depth of soil movement associated with a "natural" soil moisture regime.

– which penetrated the soil profile was observed to be physically insensitive to the various soils in which it was used. However, it was noted that certain design features – e.g. the depth of the base of the pile or pier – were directly related to the soil profile or in this case to the depth of the lower boundary of the B (clay) horizon.

The relationship between the pedological soil profile and intrinsic soil properties

The soils of the study area were widely divergent in characteristics including sands, silts, marls and clays. The clays were dominantly illitic, but ranged from relatively inert types (which included significant proportions of kaolinitic clays) to highly expansive types (which included modest porportions of montmorillonite).

Soil surveys which were undertaken both as detailed studies of specific areas and also as reconnaissance studies of the total area showed that a very large number of possible "soil types" existed (if the then-applicable definition of pedological soil type was accepted).

It was decided that, as a basis for differentiation of soil types for the purpose of this study, a requirement should be that quantitative evidence should be available to define: (1) the intrinsic volume change characteristics of each horizon and of the profile as a whole; and (2) the intrinsic strength characteristics of each horizon and of the soil profile as a whole.

Measurements of these intrinsic properties were made both in the field and in the laboratory. The values so obtained were used in part as the basis for justification of an engineering bias in the pedological classification which was subsequently adopted (see below).

In principle, it was required that soils possessing significantly different intrinsic physical characteristics should be readily differentiated through the system of classification and identification to be adopted. The converse was not a demand on the system, although in the interest of efficiency in application it was clearly desirable that the number of "types" with identical intrinsic properties should be a minimum.

In fact, it proved to be a simple matter to identify sensitive cases in which refined attention was required to separate one soil type from another. In all cases it was found that the usual morphological attributes of the soil profile were adequate to form the basis for separate identities.

The relationship between the pedological soil profile and the environmental control of soil response

The actual physical response of a soil is of course the consequence of an interaction between the intrinsic properties of the soil and the controls exerted over these intrinsic properties by changes within the environment (principally associated with the seasonal moisture cycle).

It was logical to assume that since most, if not all, of the soils could be regarded as products of the current natural environment, the soil profile itself could be considered as an expression of that (natural) environment. Thus it could be postulated that significant changes in the natural environment should be clearly reflected in differences of soil profile; or conversely, and perhaps more importantly, that in the absence of any differences in soil profile, the environment could be considered to be unchanged.

This assumption was tested in a series of field experiments in which the soil moisture regimes (under a typical vegetative cover – and also under an adjacent building) were studied over a number of seasons. It was found that little or no change in soil moisture occurred at any depth below that at which the concentrations of soluble salts or of readily leached soil constituents suggested such absence of change. Not only did these experiments support the concept of an association between the natural seasonal moisture regime and the pedological profile but it also appeared that those soils which differed significantly in intrinsic properties differed also in characteristic soil moisture regimes.

The concept of coupling

The above-mentioned conclusions from the general study, viz.: that different pedological soil types reflect different (and characteristic) foundation experiences; that different pedological soil types reflect (quantitatively) different intrinsic soil properties; that different pedological soil types reflect (quantitatively) different natural environmental controls over the intrinsic properties to create specific physical responses; and that the same differences of soil types serve in each case – all lead to one further conclusion, i.e., that the soil profile could serve as the basis for the quantitative extrapolation of knowledge basic to foundation design (which is based upon physical responses in the soil) or of foundation experience itself (see Aitchison, 1953).

The fact that the intrinsic properties of the soil and the natural environmental controls over these properties are coupled together and both linked to the soil profile means that each separate soil profile identity holds a key to the transmission of foundation experience.

The soil identities

In seeking to use separate soil profile identities as the basis for communication of knowledge pertinent to foundation engineering, it is necessary to optimise the complete system of such communication. It is essential not only that cognizance be taken of the nature of the soil variations and of the level of sophistication required in the definition of soil properties, but also that proper attention be given to the process of communication itself. It would serve no purpose to present data and descriptions relating to the soils in terms meaningful only to pedologists; for the real requirement is to communicate with architects, engineers, geologists, builders, bankers and often the lay public.

These overriding demands for simple communication led to the adoption of a codified classification reflecting certain Great Soil Group categories (after Stephens, 1956) with subdivisions based upon soil-engineering characteristics.

The number of soil identities so defined was kept to the absolute minimum. Thirty-three soil types were found to be adequate to cover all significant variations in this area despite an extraordinary diversity of major profile forms (see Fig.2 for details).

Soil mapping

In the early stages of this study there was a clear cut demand for a detailed map of soil occurrences. This supposed requirement followed naturally from the pragmatic interests of the users listed above. It is noteworthy, however, that this requirement was *not* met nor has there been any subsequent demand for such detail. The alternative, of course, lay in the adoption of a rational mapping unit coupled with a simplified procedure for recognition of soil identities.

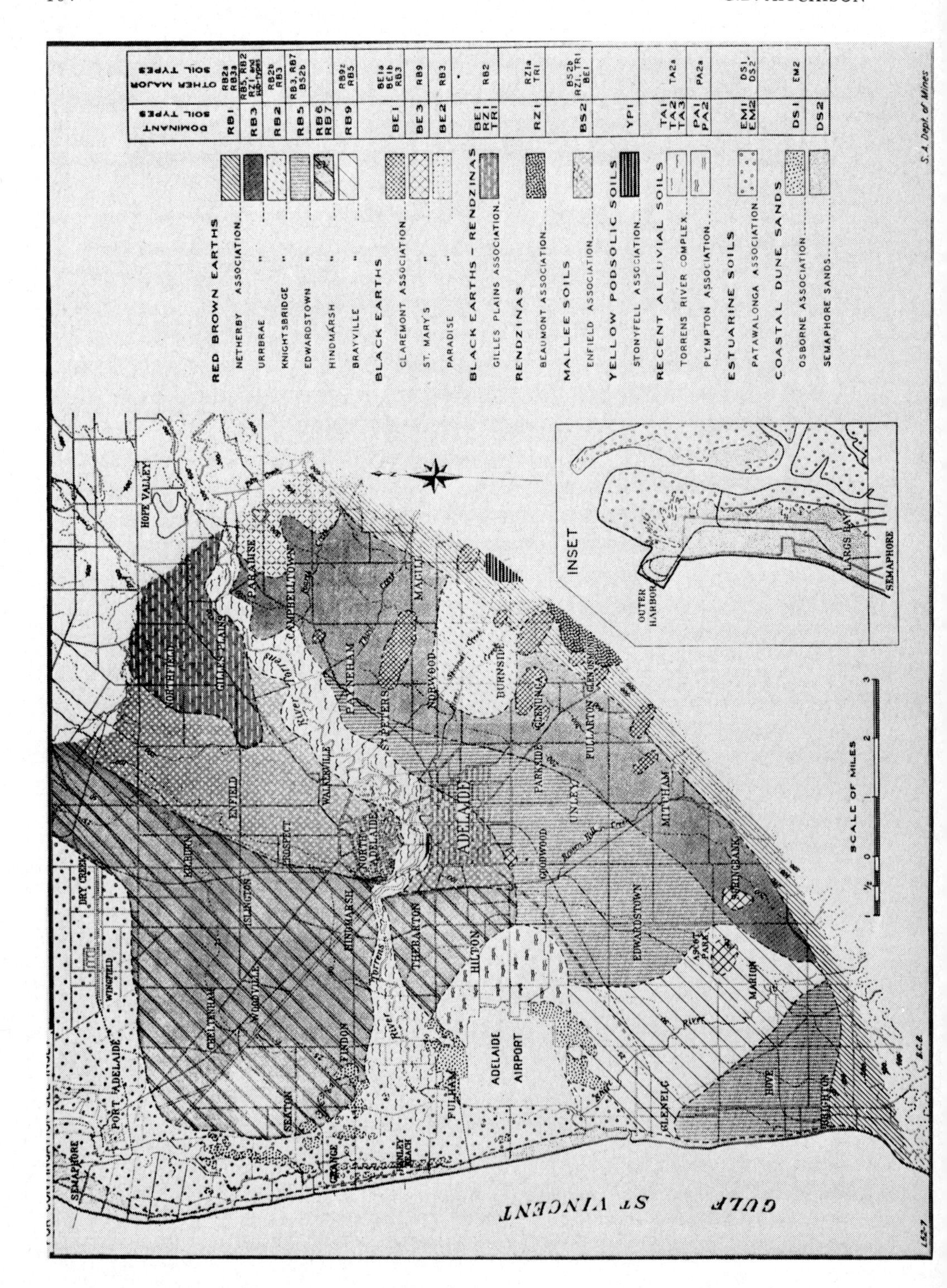
RED BROWN EARTHS
NETHERBY ASSOCIATION
URRBRAE
KNIGHTSBRIDGE
EDWARDSTOWN
HINDMARSH
BRAYVILLE
BLACK EARTHS
CLAREMONT ASSOCIATION
ST. MARY'S
PARADISE
BLACK EARTHS – RENDZINAS
GILLES PLAINS ASSOCIATION
RENDZINAS
BEAUMONT ASSOCIATION
MALLEE SOILS
ENFIELD ASSOCIATION
YELLOW PODSOLIC SOILS
STONYFELL ASSOCIATION
RECENT ALLUVIAL SOILS
TORRENS RIVER COMPLEX
PLYMPTON ASSOCIATION
ESTUARINE SOILS
PATAWALONGA ASSOCIATION
COASTAL DUNE SANDS
OSBORNE ASSOCIATION
SEMAPHORE SANDS
DOMINANT SOIL TYPES
OTHER MAJOR SOIL TYPES
INSET
OUTER HARBOR
LARGS BAY
SEMAPHORE
SCALE OF MILES
ADELAIDE
ADELAIDE AIRPORT
GULF ST VINCENT
S.A. Dept. of Mines

The map (Fig.2) involved the use of the concept of the soil association as the mapping unit. Within each soil association the number of soil types was limited so that the procedure for recognition would not be cumbersome. Eighteen associations were mapped involving a total of 33 soil types.

Presentation

The information was presented to the user in the following manner.

(1) The map (Fig.2) with the accompanying text defined the soil association within which a limited number of soil types could occur.

(2) Each soil type was described firstly by a coloured photograph of the complete profile and by further descriptions in the text.

(3) The intrinsic properties and all relevant data were tabulated for each soil type.

(4) A summary was given of the profile characteristics of each soil type, the recorded foundation experience on the soil and a suggested foundation practice to cope with the engineering properties and environmental controls (see Fig.3).

The importance of two elementary aspects of this presentation cannot be overemphasised. Firstly, there was a simply encoded descriptive term which could be easily memorised and used in discussions. Secondly, and most importantly, there was the coloured illustration of the profile presented with sufficient clarity for use by any member of the general public.

Application

In use, a series of steps was required.

(1) From the map, the soil association and the possible range of soil identities was known. (This step was not often required.)

(2) From an examination of a soil profile (preferably from an open pit but alternatively from an undisturbed continuous sequence of soil samples), the soil type was recognised by matching with the coloured photograph (and occasionally by further reference to the type description).

(3) From the tabulated data (as in Fig.3), the general nature of the foundation problems on the soil could be understood and a recommendation obtained for a suitable foundation practice (for a normal type building).

The above steps could be taken and often were taken (quite successfully) by members of the general public, as well as by professionally qualified persons.

HISTORY OF APPLICATION

In the period from 1948 to 1953 – i.e. prior to the complete publication of the results of the study – the above principles were applied as far as circumstances permitted by the research personnel concerned with the project. However, the volume of demand for infor-

FOUNDATION CHARACTERISTICS OF THE SOILS

Soil type	Characteristics of the soil profile	Foundation experience on the soil *		Suggested appropriate foundation practice	Remarks
		Incidence of foundation failure	Soil-variable related to foundation failure		
RB1	Shallow stony red-brown earth formed *in situ* on slates of the escarpment zone. Dominant type of Netherby Association †	Rare	Soil creep, *i.e.*, movement of the soil mantle downhill over a rock surface may cause some trouble	Support-foundations (piers or beams) on basement rock	High bearing-capacity permissible at proposed foundation depth
RB2	Sandy compact red-brown earth. Dominant type of Knightsbridge Association † ‡	Rare to non-existent......	—	Standard strip-footing on soil surface	Bearing capacity high. Soil movement negligible. Drainage good
RB2a ...	Similar to Type RB2, but with water-worn gravel throughout the profile	As for Type RB2	—	Strip-footing on soil surface	As for Type RB2
RB2b ...	Sandy mottled red-brown earth, characteristic of drainage lines within the Knightsbridge Association †‡	Not observed—probably not high	Some shrinkage and swelling may occur in the " B " horizon and below	Strip-footing on the soil surface (Precautions should be taken to prevent abnormal drying—as by trees—of the subsoil)	This soil occurs over a limited area only. No reliable observations of foundation behaviour are available. These remarks on foundation practices are tentative only
RB3	Red-brown earth, with heavy-textured coarse-structure " B " horizon. Dominant type of the Urrbrae Association †‡§	Common—More than 50 per cent of houses show some cracking ; many are seriously disfigured	Shrinkage and swelling movements of large magnitudes occur in the " B " horizon and below. Consequent vertical movement of surface of soil approximately 1½in. seasonally	" Pier and beam " with piers effectively supported at depths below zone of significant soil movement, *i.e.*, below 8ft. Beams to be clear of ground by at least amount of seasonal movement, *i.e.*, 1½in.	Bearing capacity high at foundation depth. Uplift force on the pier may be large enough to necessitate reinforcement in the pier shaft
RB3a ...	Similar to RB3, but with medium to large amounts of water-worn or sub-angular stone throughout the profile	Infrequent to moderate (cracking rarely serious)	As in Type RB3 shrinkage and swelling movements tend to occur within the clay horizons. However if the profile is sufficiently stony actual soil movements may be diminished by this rigid skeleton	Surface strip-footings often successful. Practice as for Type RB3 recommended where soils are not pronouncedly stony	Bearing capacity moderately high. On steep slopes precautions against soil creep should be taken
RB3b ...	Similar to Type RB3, but with deep " A " horizon (18in. or more) ‡	Infrequent (failures rarely serious)	Total soil movements as well as differential movements may be reduced much below the values for Type RB3 as a result of the additional depth of non-swelling soil material	Strip-footings on the soil surface are often successful although not recommended except on the grounds of economy. Pier and beam practices as for Type RB3 are safer	If surface strip-footings are used, extra attention should be paid to surface and sub-surface drainage of the site to minimize the wetting and drying of the subsoil
RB5	Red-brown earth, with heavy-textured fine-structured " B " horizon. Dominant type of the Edwardstown Association †‡§	Infrequent (about 10 per cent of houses show mild cracks)	Probably some small shrinkage and swelling movements occur. Actual magnitude of seasonal movement of soil not known	Surface strip-footing of increased rigidity (deep beams or inverted T-beams have been successful)	Foundation conditions tend to improve in transitional types between RB5 and RB7 and also in the travertinized soils within the Edwardstown Association nearer the sea than the modal RB5. Foundation conditions deteriorate in the transitional soils between Types RB5 and RB3. In such cases the practice for Type RB3 must be adopted

* In houses of normal construction.

Fig.3. The method of presentation of foundation experience based upon the principle of *coupled* soil types and foundation characteristics.

FOUNDATION CHARACTERISTICS OF THE SOILS—*continued*

Soil type	Characteristics of the soil profile	Foundation experience on the soil *		Suggested appropriate foundation practice	Remarks
		Incidence of foundation failure	Soil-variable related to foundation failure		
RB6	Shallow sandy saline red-brown earth, with shallow water-table. The dominant member of lower portions of the Hindmarsh Association	Infrequent	Not established. Possibly differential settlement of soil due to uneven loading (only in areas of high water-table)	Strip-footing on soil surface	Bearing capacity of soil not high, but adequate for domestic building if normal conservative practices are followed. In industrial buildings problems of bearing capacity and settlement of the soil may arise
RB7	Deep sandy red-brown earth. Slightly saline with water-table at intermediate depths. A major member of the Hindmarsh Association	Rare to infrequent	—	Strip-footing on soil surface	Bearing capacity inferior to that of Types RB3 and RB5, but adequate for domestic buildings
RB9	Degraded red-brown earth, with dull mottled surface horizons. Water-table at shallow depth. Dominant member of the Brayville Association	Infrequent to moderate ...	Shrinkage and swelling movements of small magnitude	Strip-footing of moderate rigidity on soil surface (deep beam or inverted T-beam type)	Bearing capacity not high, but adequate for suggested foundation practices
RB9a ...	Degraded red-brown earth. Transitional sub-type between Types RB5 and RB9	Moderate to infrequent ...	Shrinkage and swelling movements of small magnitude	As for Type RB9	As for Type RB9
RB9c ...	Degraded red-brown earth with heavy-textured dull-coloured horizons	Moderate to frequent	Shrinkage and swelling movements of small magnitude	Rigid strip-footing (deep beam or inverted T-beam) supported 1ft. to 2ft. below the soil surface	This soil presents a complicated foundation problem. The soil movements may be too high for strip-footings to be satisfactory, whereas the bearing capacity may be too low for the economical adoption of pier and beam practices. The suggested practice represents a compromise which could be satisfactory
RB9z ...	Degraded red-brown earth, with dull-coloured saline profile and shallow water-table. Occurs within the Brayville Association	Moderately frequent (cracking not severe)	Shrinkage and swelling movements of small magnitude	Rigid strip-footing (deep beam or inverted T-beam) supported 1ft to 2ft. below soil surface	Bearing capacity not high, but adequate for suggested foundation practices
BE1 (also BE1a and BE1b)	Black earth, with considerable profile variability. Clay from surface down to considerable depth. Normally found in escarpment outwash apron. Dominant member of the Claremont Association	Common—Practically all houses erected on this soil with normal surface foundations fail badly	Shrinkage and swelling movements of considerable magnitude, (seasonal vertical movement of the soil surface is of the order of 3in.). Differential movements may occur due to soil variability	Pier and beam, with piers effectively supported at depths below zone of significant soil movement, *i.e.*, below 7ft. Beams to be clear of ground by amount of seasonal movement, *i.e*, 3in.	Bearing capacity moderate to high at foundation depth. Uplift forces in the pier may be large enough to warrant reinforcement of the pier shaft
BE3	Black earth, with some mottling in the upper horizons. Dominant type of the St. Marys Association	Common—But severity of cracking less than in Type BE1	Shrinkage and swelling movements of appreciable magnitude (not measured)	Pier and beam or particularly rigid deep-beam or inverted T-beam types	Bearing capacity not high, therefore pier and beam constructions may not be economical

* In houses of normal construction.

FOUNDATION CHARACTERISTICS OF THE SOILS—*continued*

Soil type	Characteristics of the soil profile	Foundation experience on the soil *		Suggested appropriate foundation practice	Remarks
		Incidence of foundation failure	Soil-variable related to foundation failure		
BE2	Black earth, with some admixture of silicious sand. Dominant type of the Paradise Association	As for Type BE1	As for Type BE1	As for Type BE1	As for Type BE1
RZ1	Rendzina—Shallow dark-coloured clay soil overlying parent limestone. Dominant type of the Beaumont Association	Infrequent	Some shrinkage and swelling of the surface soil. Bearing capacity of the surface soil probably not high. Soil creep may be troublesome on steeper slopes	Support foundation (piers or beams) on basement rock (or solid zone within weathering rock)	Moderate to high bearing-capacity permissible at foundation depth
RZ1a ...	Deep dark-coloured clay soil overlying parent limestone	Moderately frequent	Shrinkage and swelling movements may be quite significant. Soil creep also may occur	As for RZ1. Pier and beam foundation is usually adopted	As for RZ1
TR1	Terra Rossa—Shallow red-coloured clay soil over lying parent limestone. Occurs with Type RZ1 in the Beaumont Association	Infrequent	Some shrinkage and swelling of the soil horizons. Soil creep may occur	Support foundation (piers or beams) on basement rock or solid zone within weathering rock	Moderate to high bearing-capacity permissible at foundation depth
BS2	Mallee soil, with lime horizon at shallow to moderate depth. A major soil of the Enfield Association	Rare to infrequent	As for Type BS2b	Strip-footings on the soil surface	As for Type BS2b
BS2b ...	Mallee Soil—Deep sandy surface soil overlying a rubbly lime horizon. A member soil of the Enfield Association	Infrequent	When subjected to flooding some settlement of the soil occurs. Differential soil movements may thus occur if part only of the foundation area is submerged. Damage is of a minor nature only	Strip-footings on the soil surface	All known foundation failures have been the result of mismanagement. Floodings due to inadequate surface drainage or to over-watering of gardens have contributed to these troubles
YP1	Sandy surface soil over-lying yellow clay above sandstone or quartzite. Dominant member of the Stonyfell Association	Rare	In winter the A_2 horizon of the soil—which is a sand with clay—becomes saturated. This horizon can become unstable if lateral support removed (as by trenching). On steeper slopes some soil creep may occur	The topographic environment of these soils may favour excavations to basement rock. On the flatter slopes the soils are sufficiently stable and free from seasonal effects to permit the use of surface strip-footings	These soils are unimportant within the mapped areas, but occur more widely in the hills districts
TA1	Gravel and sand deposits of the River Torrens	None	—	Strip-footings on the soil surface	Few buildings will be erected on these soils except in old river terraces
TA2	Dark-coloured fine-textured alluvium of the River Torrens	Not observed	Soil probably subject to settlement under load	Rigid strip-footings (of inverted T-beam type)	Buildings are rarely erected on these soils
TA2a ...	Dark-coloured alluvium from River Torrens, overlying buried red-brown earth profile (Type RB3)	Not observed	Soils may behave in similar fashion to Type RB3 or to Type RB3b	Due to variability of deposition of the alluvium, strip-footings may be unreliable. Pier and beam foundations as for Type RB3 are advisable	—

* In houses of normal construction.

Fig.3. (For legend see p. 106.)

FOUNDATION CHARACTERISTICS OF THE SOILS—*continued*

Soil type	Characteristics of the soil profile	Foundation experience on the soil *		Suggested appropriate foundation practice	Remarks
		Incidence of foundation failure	Soil-variable related to foundation failure		
PA1	Grey heavy-textured silty clay. (No profile development). A member soil of the Plympton Association	Moderate	Shrinkage and swelling and/or settlement under load	Rigid strip-footing of adequate bearing-area (inverted T-beam type) or pier and beam type with enlarged base to the piers	In some areas of these soils a foundation problem can be created by improvements to the drainage. As the water-table is lowered by drainage, shrinking of the soil may occur. Differential foundation movements may possibly result
PA2	Dark-grey heavy alluvium, overlying buried red-brown earth	Frequent................	Shrinkage and swelling movements and/or some settlement under load	As for Type PA1	These soils have affinities with the black earths. Hence foundation practices for, say Type BE1 tend to be applicable with some modifications due to the less marked seasonal moisture-changes of Types PA1, PA2, and PA2a
PA2a ...	Dark-grey heavy alluvium overlying buried degraded red-brown earth	Frequent................	As for Type PA2	As for Type PA1	
EM1	A soil composed of river alluvium overlying littoral sands. A member type of the variable Patawalonga Association	Infrequent	The soil may be sensitive to disturbance (as severe vibration) when the water-table is high; otherwise quite stable in the lower layers	Strip-footings supported about 1ft. below the soil surface	This type represents about the best foundation condition within the Patawalonga Association
EM2	A soil composed of estuarine deposits. A common member of the variable Patawalonga Association	Moderate to frequent	This soil is quite soft and may settle under building loads	Strip-footings of adequate bearing-area with proper distribution of load to avoid tendencies for differential settlement	This type is quite normal within the Patawalonga Association. The foundation problem—that of settlement under small load—is common to all of the soft estuarine deposits
DS1	Reddish-coloured dune sands of the "Osborne" coast-line. Little or no profile development. Dominant member of the Osborne Association	Rare	Some of these sands are in a loose or unconsolidated condition. Severe vibration may cause settlement; otherwise these soils are completely stable	Strip-footings on the soil surface	Where buildings are to be erected on any loose sands in an area subject to vibration, pre-consolidation of the foundation sub-strata can be achieved by flooding and vigorous stirring
DS2	White siliceous dune sands of the southern coast-line. The Semaphore Sands	Rare	As for Type DS1	As for Type DS1	As for Type DS1

* In houses of normal construction.

mation on the foundation characteristics of the soils of the Adelaide region was so great that special measures were required to cope with requests. After the publication of the study results, a small branch of the State Government Department of Mines was established specifically for this purpose of communicating information. Thereafter the demand expanded continuously so that, in due course, the field of foundation engineering via this mode of communication became a profitable one for consultants.

As the urban areas of the Adelaide plains extended with population growth (to almost double that of the pre-study numbers), further areas were mapped under the aegis of the S.A. Department of Mines. In the mapping of these new areas the identical principles were adopted and in fact many of the soils defined in the earlier studies were included in the new soil associations.

The continuity of application of this principle of coupled pedological identities and foundation engineering recommendations – over a period of up to 25 years – has been little short of astounding. Although it is impossible to obtain complete data on the number of sites examined in this period, an estimate of the order of 10,000 would not be unreasonable. (Approximately 50,000 houses have been constructed in this area in the period under review and in most cases a foundation report has been a requirement before approval to construct could be given by the relevant local government authority or before finance could be authorised by banks or other financial agencies. The majority of such reports have been based upon the principles enunciated above.)

It is, of course, gratifying to observe that research effort of this type appears to fulfil the desired objectives. Little change has been demanded by the users other than to take cognizance of extending areas (and of any new soils) and of new building or foundation techniques. Within the study area no new soil types have been required and no deletions have been reported as desirable.

AUTHOR'S COMMENTARY

It is, nevertheless, an embarrassing necessity for this author, as the original proponent of the principle of coupling of the pedological identity to the communication of foundation engineering experience, now to suggest the desirability of caution and indeed to indicate the importance of adopting – in certain circumstances – a somewhat different principle. The need for this new thinking has become obvious from experiences which have demonstrated that the environmental controls (over intrinsic properties to produce physical responses) in a totally urbanised environment are not always related (and indeed in exceptional circumstances may be totally unrelated) to those environmental controls which exist in a naturally vegetated area (and which are closely paralleled in an open-planned residential area with fully established gardens).

One early experience of this type involved the unsatisfactory performance of a block of flats erected on a black earth (designated BE1). All prior knowledge – as coupled to this soil type – suggested that foundation movements (in a normal environment) would not occur below a depth of 2.5 m. In actual fact the footings (or bored piles) were carried down to a greater depth (more than 4 m), but nevertheless the structure suffered serious distortions due to differential foundation movements of between 5 and 10 cm. As far as could be ascertained, these movements could be attributed to the modified moisture regime (at depths below 4 m) arising from increased water entry due to local concentrations of surface or drainage waters and to decreased water removal due to the relative absence of vegetation.

A totally different type of inadequacy of the coupled soil-type foundation-practice system was observed to arise when excessive water entry (from broken or inadequate agricultural drains or from the exposure by surface grading of a permeable layer (and hence a potential aquifer)) causing leaching to such an extent that changes in the solute suction occurred. Volume increases due to solute suction decreases at zero matrix suctions represented a new form of response not accounted for in the coupled pedological soil-type foundation-experience system (see Aitchison et al., 1973).

All of the foundation-engineering implications of the various departures from the preceding rules of *coupled* behaviour can be quantified and handled in the language of soil mechanics if indeed the decision is taken to *uncouple* the intrinsic properties and the environmental controls. Both of these values, i.e. the intrinsic properties and the environmental controls, can be quantified and the soil response can be computed, e.g.:

$$\Delta = \int_0^{Z_m} I_p'' \cdot N'' \cdot dZ$$

were Δ is the total vertical dimension change; I_p'' is the instability index (a measure of the stress-deformation response of the soil); N'' is the environmental control (a measure of the suction change in the soil); and Z_m is the depth of soil affected by the environmental control.

Values of the instability index I_p'' express the intrinsic properties of the soil in situ, and can be measured using appropriate laboratory techniques on undisturbed samples. Values of the environmental control N'' express the suction range to which the soil will be subjected and consequently it is through the varying values of N'' representing on the one hand the "natural" moisture regime and on the other hand the "artificial" or "urbanised" moisture regime that different patterns of behaviour of the one soil can be defined. The principles involved in the prediction of the values of N'' follow those of basic soil physics. The estimation of the depth Z_m of soil influenced by a particular soil-moisture regime also accords with computations from basic soil-physics principles but accurate input data for such computations are rarely available. In most localities it is not difficult to estimate a maximum value of Z_m, since factors such as the presence of a non-swelling layer ($I_p'' = 0$) or a permanent ground water table may be obvious determinants.

The whole of the technology required to implement the above approach has been described (Aitchison et al., 1973). Details are irrelevant in this paper.

It is important, however, that note should be taken of the compatibility in practice of the *coupled* pedological soil-type foundation-practice system and of the *uncoupled* specific computation method involving quantified values of instability indices and environmental controls.

As a first guide to foundation behaviour and desirable foundation practice the *coupled* system following the pedological soil types and data as in Fig.3 may be extremely valuable. A modified form of Fig.3 expressing the susceptibility of the soil layers at greater depth could be useful. Whenever there could be reason to anticipate a moisture regime other than the "natural" one, the concept of coupling should be discarded and computations made as suggested above. Approximate values of instability indices can be associated with the various horizons of a soil type (or alternatively the real values can be measured); each anticipated "artificial" moisture regime can be introduced in terms firstly of the relevant "environmental control (N'')" – affecting the relevant horizon – and secondly of the relevant depth of influence Z_m; and the consequential deformation Δ of the soil can be computed. It is then a task of the building designer to match structural characteristics, and details of footings with the predicted pattern of soil behaviour.

REFERENCES

Aitchison, G.D., 1953. Soil morphology and foundation engineering. *Proc. Int. Conf. Soil Mech. Found. Eng., 3rd,* 1:3–7.

Aitchison, G.D., Peter, P. and Martin, R., 1973. The quantitative description of the stress deformation behaviour of expansive soils. *Proc. Int. Conf. Expansive Soils, 3rd., Haifa,* Papers 1–8.

Aitchison, G.D., Sprigg, R.C. and Cochrane, G.W., 1954. The soils and geology of Adelaide and suburbs. *Dept. Mines, Geol. Surv. S. Aust., Bull.,* 32: 130 pp.

Stephens, C.G., 1956. *A Manual of Australian Soils.* C.S.I.R.O., Melbourne, 2nd. ed., 1962: 3rd ed., 54 pp.

Geoderma, 10 (1973) 113–122.

APPLICATION OF SOIL SURVEYS TO SELECTION OF SITES FOR ON-SITE DISPOSAL OF LIQUID HOUSEHOLD WASTES*

M.T. BEATTY and J. BOUMA

Division of Economic and Environmental Development, University of Wisconsin, Madison, Wisc. (U.S.A.)

(Accepted for publication August 16, 1973)

ABSTRACT

Beatty, M.T. and Bouma, J., 1973. Application of soil surveys to selection of sites for on-site disposal of liquid household wastes. *Geoderma*, 10: 113–122.

Safe disposal of liquid household wastes in unsewered areas is often accomplished by using a septic tank followed by subsurface soil disposal of effluent in a seepage bed. Many soils are unsuitable for this type of on-site disposal because of low permeability, close proximity to high groundwater or shallow bedrock. Soil maps are a useful tool for delineating potential problem areas and thus have come into use for purposes of selecting and evaluating building sites and for local and regional land use planning. Soil survey methodology is a useful complement to conventional site evaluation techniques. Innovative technology for on-site liquid waste disposal may be applicable for many sites that are unsuitable for traditional systems. Soil surveys are then quite useful to indicate the extent of areas in which the new technology may be applied. However, incorporation of soil information into such innovative systems will usually require a more detailed in-situ characterization of hydraulic characteristics than is currently provided by soil surveys. Successful implementation of the new technology and new uses for soil information requires cooperative interdisciplinary efforts among local, state and national health officials, soil scientists, governmental officials who develop and enforce laws and regulations, extension educators, engineers, research scientists, plumbers and home-owners. Such integrated efforts can bring about significant changes in the on-site disposal of liquid household wastes.

INTRODUCTION

Disposal of liquid household wastes in ways which protect public and private health and at the same time preserve or restore quality of water, air and land resources is an important societal consideration. Growing overall populations, increasing sizes of urban centers and generally increasing per capita production of liquid wastes compound the problem.

Public sewerage systems for collection, treatment and disposal of liquid household wastes have been developed as one response to these concerns. Such systems do not commonly embrace the burgeoning areas of low-density urban decentralization which are

---

*Contribution of University of Wisconsin-Extension Division of Economic and Environmental Development, Geological and Natural History Survey and University of Wisconsin-Madison Department of Soil Science. Based in part on work of the Small Scale Waste Management Project funded by the Upper Great Lakes Regional Commission and the State of Wisconsin.

growing around large cities of developed nations. Neither can they keep pace with the growth of more densely populated urban centers in many of the developing nations. Concerns grow for improved handling of household wastes by farmers and other residents of rural areas as well. Therefore, on-site disposal of such wastes is likely to continue to be an important means of preventing and/or solving health and environmental problems.

The design, function and performance of systems for on-site disposal of individual household wastes have not received the same attention from scientists, engineers and health officials as that given to large collective (sewerage) systems which collect and treat raw sewage and dispose of the treated products to water or land.

TYPICAL ON-SITE WASTE DISPOSAL SYSTEMS

There is an enormous variety of on-site waste disposal systems. They range from simple pipes which discharge wastes to the ground surface through unlined pits or cesspools to systems which may include partial anaerobic or aerobic treatment of wastes in a water-tight tank before it is discharged to a bed or field for absorption by the soil. The septic tank system is a common one which has been partially standardized and described (U.S. Public Health Service, 1967).

A septic tank system is made up of two components: the septic tank, of steel or concrete, used to provide partial treatment of the raw waste, and the soil absorption field where final disposal of the liquid discharged from the septic tank takes place.

The primary purpose of the septic tank is to protect the soil absorption field from becoming clogged by solids suspended in the raw wastewater. The wastewater is discharged from the home directly into the tank where it is retained for a day or more. During this time, the larger solids settle to the bottom where a sludge blanket develops, while the greases, oils, and other floating particles rise to the top to form a scum layer.

In addition to acting as a settling chamber and providing storage for the sludge and scum, the septic tank also anaerobically digests or breaks down the waste solids. Anaerobic bacteria feed on the sludge reducing its volume. In the process soluble organic matter is released from the sludge into the effluent. Methane and carbon dioxide gases are also produced and are vented from the tank through the house vent. Only about 40% of the sludge volume is eliminated in this manner, however, and about once every two to three years it is necessary to pump the tank to remove the accumulated solids. If this is not done, the tank will fill to a point where the settled solids will be resuspended and washed out onto absorption fields to clog the soil pores.

The effluent produced by a septic tank is not of high quality nor is it consistent, but this is not necessary if suitable soil is used for final subsurface disposal. Up to 60% of the BOD (Biological Oxygen Demand) and 70% of the suspended solids are removed through the tank. Significant numbers of indicator organisms, microorganisms which indicate the possible presence of disease bacteria, may be present in septic tank effluent.

The clarified liquid, or effluent, flows from the septic tank to the soil absorption field for final disposal. The field is usually a series of trenches 2–3 ft. deep. In each trench, perforated pipe overlying 12 inches of gravel distributes the liquid throughout the field. In some cases a bed of gravel and tile is substituted for the trench field. The septic tank effluent is filtered as it percolates through the soil and in a properly operating system is purified before reaching the groundwater.

A properly operating soil absorption field can treat and nearly completely purify septic tank effluent. The soil is very effective in removing BOD, phosphorus, pathogenic bacteria and viruses. Nitrogen freely moves through the system, but only if it is oxidized to NO_3^{-1} (nitrate) in well-aerated soil.

FAILURES IN SOIL ABSORPTION SYSTEMS

The typical home-owner regards a soil absorption system as satisfactory as long as it can receive all the wastewater being generated without overflowing in the house itself (hydraulic failure). This pragmatic definition needs to be extended if all major types of failure are to be included. Soil absorption systems which are adequate hydraulically may fail by: (a) delivering excessive numbers of potentially pathogenic bacteria and/or viruses to private or public water supplies; (b) causing increased nitrogen and phosphorus inputs in ground or surface water supplies, which may in turn enhance eutrophication of surface water or cause potentially toxic nitrate levels in drinking water supplies.

Hydraulic failure occurs when the soil surrounding an absorption system can not receive the liquid as fast as it is generated. Such failure is common in new systems which are too small or are poorly installed. Hydraulic failure may develop in systems which were initially adequate due to clogging or crusting which strongly reduces the infiltration rate into the soil surrounding the absorption field (McGauhey and Krone, 1967; Bouma et al., 1972).

Failures due to inadequate removal of pathogens, nitrogen and phosphorus occur when there is inadequate soil (due to lack of thickness or to coarse textures) between a soil absorption system and highly porous material such as creviced bedrock or gravel (Bouma et al., 1972). Failure due to excessive build-up of nitrogen or phosphorus in ground and surface waters may occur when there are too many homesites per unit area, even though the individual systems may each be functioning adequately to remove pathogens and absorb the liquid waste (Walker et al., 1973a, 1973b).

SITE SELECTION AND EVALUATION

The processes of site selection for on-site systems vary from folk knowledge and experience to various empirical systems of site testing, often codified into rules. In the United States the Public Health Service has developed a general reference manual (U.S. Public Health Service, 1967) which has guided many state, regional and local manuals or codes of practice. Codes of practice usually include a set of procedures for site evaluation.

The percolation test, which measures the rate of entry of water into a carefully prepared presoaked hole of prescribed diameter and depth, is a common requirement of site evaluation (U.S. Public Health Service, 1967). Other factors which may be included are: slope, presence or absence of groundwater, presence of bedrock and evidence of freedom from flooding (U.S. Public Health Service, 1967; Wisconsin Board of Health, 1969).

As an example, the current code for the State of Wisconsin contains the following limits for these criteria:

Percolation rate: 60 min. for 1 inch (2.5 cm) fall in water level in a prescribed hole.

Bedrock: at least 36 inches (90 cm) of soil over bedrock on 80% of lot area and 72 inches (180 cm) of soil over bedrock on remaining 20%.

Slope: varying as a function of the percolation rate between 10–20%.

Groundwater: highest estimated level of groundwater at least 36 inches (90 cm) below the bottom of the absorption trenches.

Flooding: sites must not be subject to flooding; 90% of lot area must be at least 2 ft. (60 cm) above flood elevation.

The traditional factors considered in and methods used for site evaluation have several limitations and inherent problems. The deceptively simple percolation test is subject to wide variation in practice and has only a very indirect relationship to performance of properly installed soil absorption systems (Bouma et al., 1972). The presence of a temporary groundwater table is often not obvious from traditional soil boring procedures when these are carried out in the dry season (as is most convenient for field operations). Typical procedures for site evaluation rarely convey any idea of the possible interrelationships of waste, soil, water, air and organisms at proposed disposal sites. Rather, site evaluation procedures are reduced to a series of steps, minimum criteria and tables of required areas.

CURRENT ROLES OF SOIL SURVEY

As a natural consequence of the uses of soil surveys for engineering purposes, this resource inventory has been used as a tool for evaluation of potential sites for on-site waste disposal. This effort has met with considerable acceptance because soil surveys (particularly detailed surveys) have provided information directly useful in site evaluation and complementary to information obtained by more traditional site evaluation procedures.

Soil properties such as structure and texture, as determined by pedologists, have proven useful supplements to bore hole data provided by plumbers, surveyors and engineers. Morphological characteristics associated with various water regimes in soil (natural soil drainage) have proven helpful in defining groundwater levels under some conditions. Soil maps also give useful information on general slopes and the presence of shallow bedrock.

Properties and qualities of soils studied and mapped in soil surveys have been shown to be related, at least qualitatively, to the performance of soil absorption systems. In addition to the interpretations provided in published soil surveys, various specialized publications relating performance of systems to soils are available (Mokma and Whiteside, 1973).

Fig.1 and Table I show interpretations of detailed soil survey information for site evaluation for on-site waste disposal using currently available judgements of severity of limitations, which are in turn linked to currently available technology. The included soils are listed and classified in Table I. Fine-loamy, fine-silty and loamy mixed, mesic Typic and Lithic Hapludalfs are the most extensive families and subgroups in the square mile selected for display and discussion. Typic Haplaquolls occupy extensive drainage-ways. Only two bodies (Fox and Theresa soils in the northwest corner of the area) have slight limitations and are acceptable for standard soil absorption systems using septic tanks.

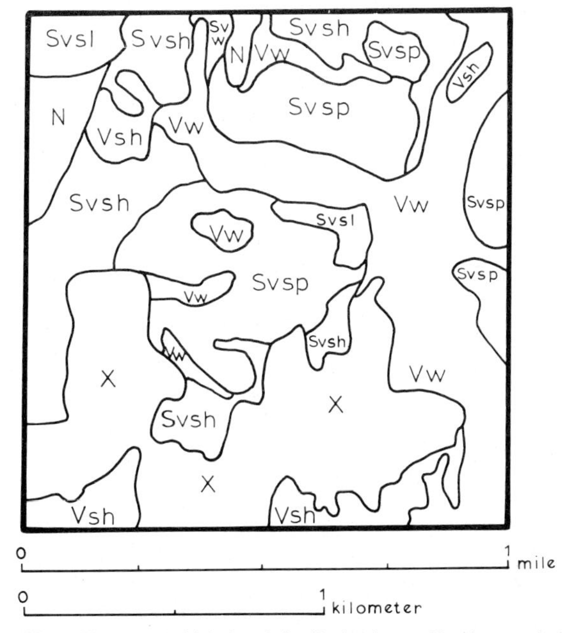

Fig.1. Degrees and kinds of site limitations affecting on-site sewage disposal systems. *N* = slight (2) (13); *Svsh* = severe, shallow, creviced bedrock (4); *Svsl* = severe, slope (3); *Svsp* = severe, slowly permeable soil (9); *Svw* = severe,wetness (6); *Vsh* = very severe, very shallow, creviced bedrock (12); *Vw* = very severe, wetness (1) (5) (7) (8) (10) (11); *X* = quarries, urban area and variable made land–not rated. (Numbers in parenthesis refer to soil series in Table I.)
Source: Soil Survey of Milwaukee and Waukesha Counties, Wisconsin, U.S.A. U.S.D.A. Soil Conservation Service and Cooperators; July, 1971; Map sheet 18 and Table 8.

Fig.1 and Table I show that other sites have severe and very severe limitations which render them unacceptable for standard systems, because soils are either too steep, too wet, too slowly permeable or too shallow over bedrock to meet current Wisconsin code criteria.

FUTURE ROLE OF SOIL SURVEY WITH AVAILABLE NEW TECHNOLOGY

Every soil to be used for on-site liquid waste disposal is, in fact, a problem soil; the problems vary in character and magnitude among different soils as was discussed in preceding sections. Safe, economical and reliable on-site liquid waste disposal can be

TABLE I

Classification and limitations for on-site sewage disposal systems of soils in a one square mile area of Waukesha County, Wisconsin*

Series (and slope phase as applicable)	Family	Subgroup	Degree of limitation
(1) Ashkum	fine, mixed, noncalcareous, mesic	Typic Haplaquolls	very severe
(2) Fox	fine-loamy, over sandy-skeletal, mixed, mesic	Typic Hapludalfs	slight
(3) Hochheim (12–20% slopes)	fine-loamy, mixed, mesic	Typic Argiudolls	severe (slight or moderate on lesser slopes)
(4) Knowles	fine-silty, mixed, mesic	Typic Hapludalfs	severe
(5) Lawson	fine-silty, mixed, mesic	Cumulic Hapludolls	very severe
(6) Matherton	fine-loamy over sandy or sandy-skeletal mixed, mesic	Udollic Ochraqualfs	severe
(7) Mequon	fine, mixed, mesic	Udollic Ochraqualfs	very severe
(8) Ogden	clayey, euic, mesic	Terric Medisaprists	very severe
(9) Ozaukee	fine, mixed, mesic	Typic Hapludalfs	severe
(10) Palms	loamy, euic, mesic	Terric Medisaprists	very severe
(11) Pella	fine-silty, mixed, noncalcareous, mesic	Typic Haplaquolls	very severe
(12) Ritchey	loamy, mixed, mesic	Lithic Hapludalfs	very severe
(13) Theresa	fine-loamy, mixed, mesic	Typic Hapludalfs	slight

*Steingraeber, J.A. and Reynolds, C.A., 1971, Soil Survey of Milwaukee and Waukesha Counties, Wisconsin; U.S. Department of Agriculture Soil Conservation Service and Cooperating Agencies, 177 pp. and maps.

achieved in many ways which are different from the conventional procedure using a septic tank and soil disposal field or bed. For example, treatment of the raw wastes can be improved by using mechanical aeration units or by disinfection, and soil disposal and purification can be improved by using innovative seepage-bed designs and systems of liquid application (Bouma et al., 1972; Bouma et al., 1973a, b; Goldstein, 1973). In fact, the technology involved has been developed sufficiently to conclude that satisfactory on-site liquid waste disposal can be achieved almost anywhere if sufficient time, money and effort are expended. However, economic and health considerations, relating to construction and maintenance, severely limit the practicality of many complex systems. Use of soil, which is very efficient and cheap as a "biologic filter", remains therefore very attractive in any disposal system.

Innovative disposal systems in soils will have to be designed specifically for different soil and site conditions. For example, insufficient soil is available for percolation in thin loamy soils over creviced bedrock; whereas, the problem in slowly permeable soils is caused by the restricted infiltration into the absorption field. In the case of shallow soils additional soil has to be added to the soil in situ, forming a mound system (Bouma et al., 1972), and mounds or large subsurface seepage areas are required in the latter case to ensure adequate infiltration. When either natural soil in situ or soil material as a fill is to be used for on-site waste disposal there is a need for more specific knowledge about their potential physical behavior than can be obtained with the currently used soil percolation test which is physically poorly defined (Bouma, 1971) or with a qualitative soil drainage classification (Bouma, 1973). Physical methods are available now to measure relevant soil physical characteristics in situ (Bouma et al., 1972; Klute, 1972) and these methods have to be applied before the potential role of any soil in an innovative disposal treatment system can be fully assessed.

Future use of soil maps to help define the potentials of sites for application of advance technology to on-site waste disposal will be likely to require more complete data on physical properties of soil which affect hydraulic conductivity at or near saturation. In addition to aiding in site selection these data can be used to determine the design and construction of innovative systems.

Assuming the availability of these data, Fig.2 shows portions of the square-mile area shown in Fig.1 where soil and site conditions are potentially acceptable for either standard soil absorption systems or systems based on new, innovative technology and more thorough soil characterization. In addition to the bodies of Fox and Theresa soils which are suitable for standard soil absorption systems, the soil map may be interpreted to identify areas which are potentially suitable for innovative, specialized treatment and disposal systems such as the mound system and innovative seepage bed designs and systems of liquid application.

As is true in interpretations of soil surveys for farming,soil with a given degree of limitation must be considered in relation to each system or level of technology associated with a given land use. In the case shown in Fig.2 limitations are so severe that the use of traditional soil absorption systems (simple technology) is geographically very limited.

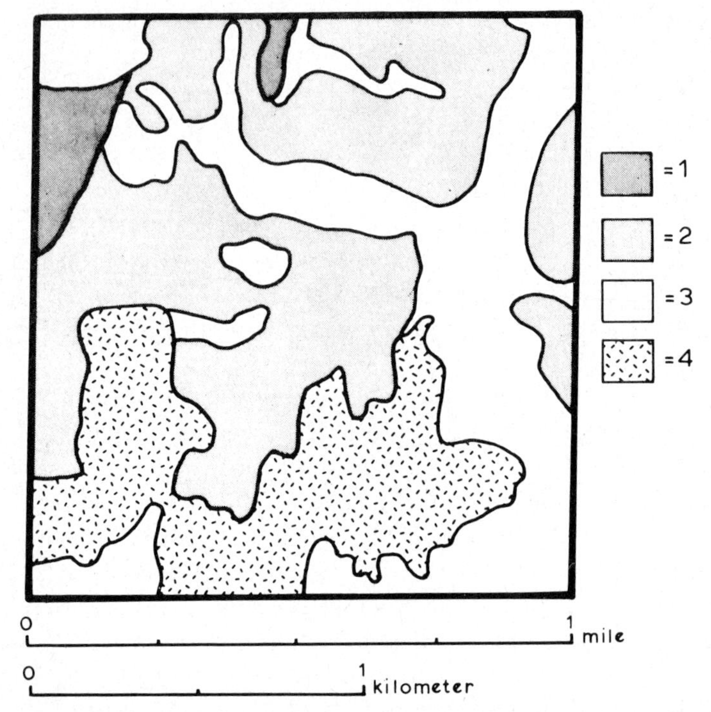

Fig.2. Areas within a one square mile tract with varying suitability for standard or specialized soil absorption systems. *1* = standard soil absorption system acceptable; *2* = specialized mounded systems potentially acceptable; standard systems not acceptable; *3* = neither standard nor specialized systems acceptable; *4* = quarries, urban area and variable made land–not rated for soil absorption systems.

More advanced technology may overcome certain of the limitations so that on-site waste disposal systems may operate successfully on a considerably larger part of the area.

The processes of interdisciplinary research field observations and field testing of new technology needed to relate soil mapping units to farming systems are equally necessary to the development of sound new predictions of soil performance for on-site waste disposal.

NECESSARY ASSOCIATED ACTIVITIES

Introducing and perfecting the uses of soil survey in site selection for on-site waste disposal systems requires carefully planned and sustained attention to several points. First, is development of mutual technical understandings among engineers, health sanitarians, state and local code administrators and soil scientists. Second, changes in codes, laws or administrative rules may be needed. This involves careful work with government officials. Third, people in construction such as plumbers, surveyors, builders, contractors and real estate developers must become fully informed of direct opportunities and benefits resulting from use of soil surveys. Fourth, persons planning to buy or build homes to be served by such systems need to know about the role of soil survey in site selection and design as a part of broader general knowledge on buying and building rural homes.

The careful matching of attention given to the technological and the educational aspects of a program of change which introduces soil surveys into site selection processes is essential if useful new ideas and new information are to be accepted and diffused rapidly among the various groups described.

Beatty and Lee (1972) have described a model for relating information from soil science to land use planners. It involves carefully planned initial contacts, sustained interdisciplinary and interagency follow-up, mutual development of research needs and development of a "second generation" of data, procedures and mutual understandings which lead to fruitful continued programs. The same principles apply to the topic of this paper. Programs for bringing together professionals must be accompanied by continued education programs for local officials, tradesmen, homebuyers and the general public. Such educational programs are analogous to those used to diffuse new information on soil science among multiple audiences (e.g., farmers, local agricultural technicians, fertilizer dealers, etc.) in agriculture and agri-business.

An active program of this kind in the state of Wisconsin has produced significant results. Several thousand lots proposed for homesite development are rejected each year on the basis of pedological criteria. Some regional land use planning programs reflect soil survey inputs strongly (Southeastern Wisconsin Regional Planning Commission, 1969).

Local sanitary codes in the majority of Wisconsin counties have specific soil survey materials incorporated within their texts. The State Division of Health employs a soil scientist (Ph. D.) on its program staff. Training programs for plumbers, surveyors and engineers have diffused new knowledge and upgraded the practices of many of these tradesmen and professionals. Demands for soil surveys have increased significantly as a result of these interrelated activities.

SUMMARY AND CONCLUSIONS

On the basis of approximately 15 years of experience we suggest that a program to introduce the use of soil surveys into site selection for on-site waste disposal have these major elements: (1) identification of needs and roles for soil survey information jointly with sanitarians, engineers, etc.; (2) development of initial interpretations and joint working relationships; (3) training programs; (4) full-scale program of using soil surveys; (5) research based on needs developed by the new program; (6) development of new and better information on soils, waste engineering, health, water quality, etc.; (7) re-training and upgrading of soil surveys together with laws, rules, practices of waste disposal on a continuing basis.

Such a program can produce positive results.

REFERENCES

Beatty, M.T. and Lee, G.B., 1972. Relating soil science teaching to land use planning. *J. Agron. Educ.*, 1: 61–64.

Bouma, J., 1971. Evaluation of the field percolation test and an alternative procedure to test soil potential for disposal of septic tank effluent. *Soil Sci. Soc. Am. Proc.*, 35: 871–875.

Bouma, J., Ziebell, W.A., Walker, W.G., Olcott, P.G., McCoy, E. and Hole, F.D., 1972. *Soil Absorption of Septic Tank Effluent. Information Circular No.20*. Geol. and Nat. Hist. Surv., Univ. Ext., Univ. of Wisc., Madison, Wisc., 235 pp.

Bouma, J., 1973. Use of physical methods to expand soil survey interpretations of soil drainage conditions. *Soil Sci. Soc. Am. Proc.*, 37: 413–421.

Bouma, J., Converse, J.C. and Magdoff, F.R., 1973a. *Dosing and Resting to Improve Soil Absorption Beds*. Paper No.73-25b, presented at ASAE meetings in Lexington, Ky.

Bouma, J., Converse, J.C. and Magdoff, F.R., 1973b. A mound system for disposal of septic tank effluent in shallow soils over creviced bedrock. *Symp. Int. Conf. Land Waste Management, Ottawa.*

Goldstein, S.N. and Moberg, W.J., 1973. *Wastewater Treatment Systems for Rural Communities.* Commission on Rural Water, Washington, D.C., 340 pp.

Klute, A., 1972. The determination of the hydraulic conductivity and diffusivity of unsaturated soils. *Soil Sci.*, 113: 264–277.

McGauhey, P.H. and Krone, R.B., 1967. Soil mantle as a waste water treatment system. *Sanitary Eng. Res. Lab. Rept., Univ. California, Berkeley*, No.67-11: 200 pp.

Mokma, D.L. and Whiteside, E.P., 1973. The performance of septic tank disposal fields in representative Michigan soils. *Mich. Agric. Exp. Sta. Res. Rept.*, 157 pp.

Southeastern Wisconsin Regional Planning Commission, 1969. *Soils Development Guide. Planning Guide No.6*. Southeastern Wisconsin Regional Planning Commission, Old Ct. House, Waukesha, Wisc., 247 pp.

U.S. Public Health Service, U.S.D.H.E.W., 1967. *Manual of Septic Tank Practice*. U.S. Gov. Printing Office, Washington, D.C., Pub.526, 93 pp.

Walker, W.G., Bouma, J., Keeney, D.R. and Magdoff, F.R., 1973a. Nitrogen transformations during subsurface disposal of septic tank effluent in sands, I. Soil transformations. *J. Environ. Qual.*, in press.

Walker, W.G., Bouma, J., Keeney, D.R. and Olcott, P.C., 1973b. Nitrogen transformations during subsurface disposal of septic tank effluent in sands, II. Groundwater quality. *J. Environ. Qual.*, in press.

Wisconsin State Board of Health, 1969. *Private Domestic Sewage Treatment and Disposal Systems.* Wisconsin Administrative Code. Wisconsin Department of Health and Social Services, Madison, Wisc., 16 pp.

Geoderma, 10 (1973) 123–130

THE PROPERTIES OF NEW ZEALAND SOILS IN RELATION TO EFFLUENT DISPOSAL

N. WELLS

Soil Bureau, Department of Scientific and Industrial Research, Lower Hutt (New Zealand)

(Accepted for publication August 16, 1973)

ABSTRACT

Wells, N., 1973. The properties of New Zealand soils in relation to effluent disposal. *Geoderma*, 10: 123–130.

Four site characteristics and seven soil properties are discussed in relation to the disposal of effluents. The properties that limit the assimilation of effluents are listed in terms of the soil criteria used in New Zealand. Systems of effluent disposal are considered for different classes of slope and for different volumes of effluent. Soil groups are rated for their capacity to cope with agricultural effluents. Yellow-brown loams (Andosols) have the ideal combination of properties for effluent disposal.

INTRODUCTION

The soil scientist is being increasingly called upon to advise on problems created by effluents. Small-scale dispersed production units are being concentrated on the grounds of efficiency and effluent volumes are increasing because of the greater water use required for higher standards of product hygiene. At the same time deterioration of ground-water quality is a public issue and these factors combined necessitate a better assessment of the capacity of soils to handle effluents in terms of the amount of water and its loading.

The disposal of effluents into soils can be limited by the four site characteristics of drainage, slope, climate and vegetation. Water movement in the profile is influenced by the physical properties of structure, texture, consistence, porosity, pans, gleying and organic matter. These site characteristics and soil properties are considered in terms of the units used in New Zealand (Taylor and Pohlen, 1962). The effluents considered are those from agricultural industries, small domestic sewage schemes and farming. The problems raised by some industrial effluents may be too specialised for these generalisations.

SITE CHARACTERISTICS

Drainage

Soils of only three of the seven soil drainage classes described below are suited to effluent disposal schemes. The imperfectly and poorly drained soils will not absorb water at the rate required, while the excessively free draining soils will tend to pass nutrients and undesirable components straight through into the ground waters. If the less well drained soils have to be used a pre-treatment in anaerobic and aerobic lagoons would be required before discharge.

Very poorly drained. Water remains at or on surface for greater part of year. These soils with peaty topsoils and grey subsoils are unsuitable for effluent disposal.

Poorly drained. Water is removed so slowly that soil remains above field capacity most of the year. These conditions relate to a high water table, a slowly permeable layer and/or seepage and they render the soil unsuitable for disposal of effluent in the raw state.

Imperfectly drained. Water is removed from the soil slowly enough to keep it above field capacity for significant periods of time but not continuously. These soils commonly have a slowly permeable layer, a high water table, or additions through seepage. They are unsuitable for disposal of effluent in the raw state.

Moderately well drained. Water is removed somewhat slowly so that the soil is above field capacity at certain times. These soils may be unsuitable for effluent disposal every day of the year but are useful for disposing of small volumes.

Well drained. Water is removed readily but not rapidly from these soils. They are not above field capacity for any length of time but can retain near optimum moisture for long periods. These soils are often silt loams and are very suitable for the disposal of effluents.

Somewhat excessively drained. Water is removed rapidly so that moisture deficiency frequently limits plant growth. Many of these soils have little horizon differentiation, are sandy and very porous. These soils can be used for effluent disposal where volumes are high and nutrient contents low.

Excessively drained. Water is removed rapidly and very little is retained by the soil. These soils may be lithosols, steep or very porous and are unsuitable for effluent disposal as undesirable components will pass rapidly into ground waters.

Topography

In conventional irrigation the system is selected on economic factors related to topography and inherent soil productivity. In effluent disposal systems the costing should be built into the cost structure of the product. A greater range of topographic units can be covered under this costing or may have to be covered because of other site considerations.

Flat to undulating. Most slopes are less than 3°. Flooding from border dykes is a suitable system for large volumes of effluent containing grease, fats and semi-solid organic particles which would block piped systems (e.g. effluents from abattoirs and meat processing factories). Spray irrigation can be used for smaller volumes after pre-treatment to remove particles.

Rolling. Most slopes are under 12°. Multiple sprinklers connected to movable pipes are used on easy rolling land when the effluent is in solution or fine suspension and volumes moderate (e.g. effluents from dairy factories). High pressure single nozzles from fixed pipelines are used where the movement of pipes is difficult.

Moderately steep. Slopes are usually in the range 12°–23° but some may be up to 30°. High pressure, single outlet, rotating nozzles are used at vantage points on the land surface. The properties of the soil become very important as deterioration in vegetation cover can induce erosion.

Steep. Slopes in this unit are usually between 30° and 38°. Disposal of effluents by irrigation on these slopes presents erosion problems unless there is a strong cover of vegetation (e.g. an established forest). In steepland areas it is often possible to construct a series of lagoons which can improve the quality of the effluent up to a standard acceptable for discharge into watercourses.

Precipitous. In New Zealand this class, with slopes about 55°, covers large areas but they are in remote regions. These soils are poised on the limit for debris–avalanche and should not be disturbed.

A disposal system less dependent on topographic class is the suction tanker with sprinkler discharge. These clear out sumps and holding tanks for small volumes of liquids and thin slurries (e.g. effluents from small piggeries and from large milking sheds). Satisfactory contract arrangements require a large number of units in a limited area. The discharge from these tankers can be made on to a suitable class of land at a distance that is less critical than in piped systems. It has greater flexibility to fit changes in crop production.

Climate

The climate at a site has two main features – moisture and temperature. Irrigation of drier soils with effluent will give a greater return in plant growth for the water component of the effluent. The wetter soils are seldom at wilting point but they can benefit from the nutrient content of the effluent.

Moisture class

Subxerous. Soils below field capacity all year and below wilting point for six months or more. The brown-grey earths are in this class and irrigation is necessary for intensive agricultural production.

Subhygrous. Soils below field capacity for more than five months of the year and below wilting point for one to five months. The yellow-grey earths are in this class and irrigation of pastures is required to maintain production in summer.

Hygrous. Soils are above field capacity for most of the year and they do not reach wilting point for any month. The drier subclasses are below field capacity for one to five months while the moister subclasses are above field capacity all year. Holding tanks may be required for effluent to avoid irrigation of these soils during rain.

Hydrous. Soils above field capacity all the year and over-wet for long periods. These soils include B-gleyed podzols and some yellow-brown earths; they require drainage and are unsuitable for effluent disposal.

Temperature zones

The Alpine zone above 2,000 m represents the area too cold for plant growth. The sub-Alpine zone, sited above the main forested areas, commences at about 1,000 m in the south of New Zealand and at about 1,500 m in the north. In both these zones and in the sub-Antarctic zone the rate of organic matter decomposition is too slow for the simple disposal of effluent by irrigation. For small-scale units some form of heated holding tank could be required to semi-process the effluent from ski villages before disposal.

In the temperate zone, in the more southern parts of New Zealand, the freezing of water in pipes can limit the type of irrigation system used to dispose of effluents. However, the rate of oxidation of organic wastes is adequate in this zone to cope with organic effluents from agricultural enterprises. In the subtropic zone the more rapid rate of decomposition can require the use of the more permeable soils to prevent odour problems.

Vegetation

In New Zealand most effluents have been irrigated on to ryegrass-clover pasture. The use of effluents rich in nitrogen tends to change the composition of the sward towards a grass-dominant pasture. The rooting of plants has been weakened in pastures irrigated with effluents and turf-pulling by animals has occurred. In areas over-treated with effluents soil pugging by animal hoofs is a problem on the less suited soils and in extreme cases only deep-rooted weeds survive.

Permanent pastures have organic-rich topsoils in which the organic matter is assisting in forming stable crumb structures with good permeability. Cropped soils, however, usually have lower permeability as organic matter has been lost, aggregates weakened, and soil particles compacted by machinery operations.

Toxicity factors, such as potential pathogens, may limit the type of vegetation at a site for effluent irrigation. Closed cycles of animal – effluent – crop – same animal should be avoided. The disposal of human wastes on to soils under forest is an experiment procedure in New Zealand but the use of soils under pasture or crops for this purpose has not been studied. The application of dairy wastes to pastures has been studied by McDowall and Thomas (1961), who reported on irrigation systems, nutrient value, acidity, and temperature of the effluent. Wells and Whitton (1966, 1970) analysed white clover and grass from areas irrigated by dairy factory wastes and by meat works wastes and showed some imbalance of element composition in the plants.

SOIL PROPERTIES

The systematic profile descriptions used for soil survey in New Zealand list the properties in the following order: thickness of horizons, colour, texture, consistence, porosity, structure, organic matter and nature of horizon boundary. The following properties have been selected for their relevance to water permeability: structure, texture, consistence, porosity, pans, gleying and organic matter.

Structure

This is classed according to form, size and degree of development. The listing of forms as: prism, column, block, nut, plate, granule and crumb is only in a general way an order of increased permeability to water, as degree of development is very important. The different forms of structure can also relate to changes in permeability; a blocky structure when dry will be highly permeable via the cracks between blocks but these can seal when the profile wets up and permeability will be reduced. A good crumb structure, however, will be permanently permeable.

Texture

In general the coarser the primary particles the more permeable will be the soil and Horn (1971) has used texture to estimate permeability. However, many soils would be too permeable for effluent disposal schemes unless some cementation or compaction of the primary particles occurred in the subsoil. Loam and silt loam textures allow the most suitable rate of percolation for effluent disposal. Where the water volumes are high, but the effluent component is low, a sandy loam could be desirable. A clay loam or silty clay loam would require further qualifications, e.g. a well developed granular or nut structure, to be suitable for effluent irrigation.

Where the primary clay particles have been very strongly bonded by iron compounds to give non-plastic oxide-rich clays the term friable clay has been used to denote a loamy field texture with good permeability in spite of a high clay content by mechanical analysis.

Consistence

The consistence of a soil relates to its cohesion and adhesion at various moisture contents. Soil consistence reflects the behaviour of the soils under use at various moisture levels. Under effluent irrigation the consistence when dry (a rating from soft to hard) will seldom apply. Consistence when wet to above field capacity concerns the degrees of stickiness and plasticity. These properties are of importance if cropping is used on the areas treated with effluents and they also have a role in hoof damage on permanent pastures that have been overtreated.

Porosity

The classification of soil pores is by a combination of three factors: shape (vesicular, interstitial or tubular), size, and abundance. Consideration has to be given to voids between primary particles (e.g. between sand grains) and to voids between aggregates (e.g. cracks in structure units). Some structure units imply certain types of porosity, e.g., crumbs are porous but granules are not.

The presence of abundant pores is beneficial to effluent irrigation systems as they increase permeability and assist in aeration and oxidation of the organic component. Pores that can be seen in the field are usually of sufficient size to be freely draining in a few hours under gravity.

Laboratory measurements of macro-porosities for the soil groups in New Zealand have been published by McDonald and Birrell (1968). They showed very high values (25%) in yellow-brown pumice soils and high values (12%) in most soils derived from volcanic ashes. Low values (6%) were found in subsoils of yellow-grey earths and strongly weathered yellow-brown earths.

Pans

Several types of compact horizons can occur in soil profiles. In most cases these will impede drainage and reduce the permeability. These pans can be typed as: calcium carbonate, fragipan, clay pan, iron pan, silica pan and humus pan. Some of these features can be seen as an asset in effluent irrigation on soils that would be too permeable without a fragipan or a weakly developed iron pan or clay pan.

Gleying

Anaerobic conditions, produced by slow decomposition of organic matter in wet soils, result in greyish colouration of subsoils. These are an obvious indication of unsuitable conditions for the disposal of effluents. Gley horizons are classified according to the degree of gley mottling: weakly gleyed, moderately gleyed, strongly gleyed and very strongly gleyed.

Organic matter

Accumulations of organic matter in topsoils are an indication of high rainfall, high water table or a low rate of decomposition. Raw non-decomposed organic matter is permeable to water but highly decomposed organic matter has gel-like properties in which the movement of water is retarded. Peats are classified into ten classes of decomposition but all of these would be considered as being unsuitable for receiving raw effluents.

SOIL CLASSIFICATION AND EFFLUENT DISPOSAL

The special land-use implied by effluent disposal requires some rearrangement of the relevant soil properties noted in the pedological appraisal. Application of soil classification concepts, based on normal appraisal, can still be applied but it must be borne in mind that at a particular site the phasing on stoniness, drainage or erosion may have an overriding impact.

In Table I the common names of the New Zealand soil classification (Taylor and

TABLE I

Soil classifications and effluent disposal

New Zealand class	FAO–UNESCO class	Rating*
Brown-grey earths	Luvisols	M
Yellow-grey earths	Phaeozems, Luvisols	M
Yellow-brown earths	Cambisols, Acrisols	V
Podzols, B gleyed	Podzols	S, U
Rendzinas	Rendzinas	U
Yellow-brown pumice soils	Andosols	R
Yellow-brown loams	Andosols	R
Brown granular loams and clays	Acrisols	M
Red and Brown loams	Andosols, Ferralsols	R–M
Gley soils	Gleysols	U
Organic soils	Histosols	U
Recent soils from alluvium	Fluvisols	R
Recent soils from volcanic ash	Regosols	R
Steepland soils	Cambisols	U

*R = rapid percolation rate (10-3 cm/h approx.)
M = moderate percolation rate (3-1 cm/h approx.)
S = slow percolation rate (1-0.1 cm/h approx.)
V = highly variable
U = unsuitable

Pohlen, 1962) have been listed, together with their broadest-correspondence in the FAO–UNESCO "Soil Map of the World" system (Dudal, 1968) and rated for their capacity to cope with effluents. This has been complied from data on macro-porosity by McDonald and Birrell (1968) and from unpublished data on permeabilities. The soils having the best characteristics for this type of land-use are the yellow-brown loams (Andosols) derived from andesitic ashes in the 2,000 to 7,000 years age group. Typically these soils have well developed medium nut structures, are silt loam in texture, have good friability when moist but are non-sticky and non-swelling when wet and have considerable amounts of organic matter incorporated in the mineral soil.

CONCLUSIONS

The negative characteristics that limit soils for the disposal of raw effluents by irrigation in New Zealand can be listed under characteristics of site and properties of soil profile (see Tables II and III).

TABLE II

List of site characteristics for disposal of raw effluents

Drainage	Slope	Moisture	Temperature
Very poor*	flat to undulating	subxerous	sub-antarctic*
Poor*	rolling	subhygrous	alpine*
Imperfect*	moderately steep	hygrous	sub-alpine*
Moderate	steep*	hydrous*	temperate
Well	precipitous*		subtropic
Somewhat excessive			
Excessive*			

*Characteristics unsuited to raw effluent disposal.

TABLE III

Soil profile properties for disposal of raw effluents

Structure	Texture	Pores	Pans	Gleying
Prism*	sand*	few*	carbonate	weak*
Column*	sandy loam	many	fragi	moderate*
Block*	silt loam	abundant	clay*	strong*
Nut	loam		iron*	very strong*
Plate	silty clay loam*		silica*	
Granule	clay loam*		humus*	
Crumb	clay*			

*Properties less suited to raw effluent disposal.

REFERENCES

Dudal, R., 1968. *Definitions of Soil Units for the Soil Map of the World.F.A.O. World Soil Resour. Rep.*, *33*: 72 pp.

Horn, M.E., 1971. Estimating soil permeability rates. *J. Irrig. Drain. Div. Am. Soc. Civ. Eng.*, 97: IR2: 263–74.

McDonald, D.C. and Birrell, K.S., 1968. Physical data for modal profiles. In: *Soils of New Zealand. N.Z. Soil Bur. Bull.* 26(2): 139–141.

McDowall, F.H. and Thomas, R.H., 1961. Disposal of dairy factory wastes by spray irrigation on pasture land. *Pollut. Advis. Counc. Pub. N.Z.*, 8: 94 pp.

Taylor, N.H. and Pohlen, I.J., 1962. *Soil Survey Method. N.Z. Soil Bur. Bull.*, 25: 242 pp.

Wells, N. and Whitton, J.S., 1966. The influence of dairy factory effluents in soil and plant chemistry. *N.Z. J. Agric. Res.*, 9: 565–75.

Wells, N. and Whitton, J.S., 1970. The influence of meatworks effluents on soil and plant composition. *N.Z. J. Agric. Res.*, 13: 494–502.

Geoderma, 10 (1973) 131–139

THE USE OF SOIL SCIENCE IN SANITARY LANDFILL SELECTION AND MANAGEMENT

F. GLADE LOUGHRY

Division of Community Environmental Services, Pennsylvania Department of Environmental Resources, Harrisburg, Penn. (U.S.A.)

(Accepted for publication June 26, 1973)

ABSTRACT

Loughry, F.G., 1973. The use of soil science in sanitary landfill selection and management. *Geoderma*, 10: 131–139.

Sanitary landfill is a method of waste disposal utilizing soil in the elimination of discarded foods, fibers, and artifacts. Masses of municipal waste placed in trenches or in soil encased mounds decay in a soil environment. Soil serves as a lining for the site, as cover, as a physical and chemical renovator, and as a medium for restoring vegetation and normal land use. The soil characteristics that are effective in this process include good natural drainage, a deep moderately permeable profile, and loamy texture. These can be identified and preliminary site evaluation made with the help of detailed soil surveys.

Solid waste is always a companion of production and use of goods. The kind of waste differs if we are considering food, clothing, housing fuel, machines, or luxury items, but the principle is the same. Part of the material received by the user is not of immediate utility or wears out and is called waste. As it is discarded it becomes a problem. It increases at an alarming rate as population grows. Even greater factors in its accumulation are increasing technological complexity of industries and affluence of people. But even in primitive or low-income areas there are solid waste problems.

The kind of solid waste that has been given the most consideration in urban societies is the municipal refuse. It consists of newspapers, cardboard containers, food scraps, metals of many kinds, glass bottles, plastic containers, lawn clippings, shrub and tree prunings, debris from demolished buildings, ashes, and earth from excavations. It is collected from homes, offices, commercial establishments, institutions, and some industrial plants. It accumulates rapidly. For cities in the United States it is estimated that the per-capita daily production of this kind of waste is now about 2.5 kg (Commonwealth of Pennsylvania, 1970). The corresponding figure in 1960 was 1.8 kg. Future projections make it as much as 3.1 kg by 1982.

If allowed to accumulate for even a short time this mixture of materials putrefies, producing noxious odors and providing food and shelter for rats and roaches and a breeding place for flies. If exposed to rain or placed in water it produces a leachate that has a biochemical oxygen demand which on occasion is as much as 100 times that of raw sewage received at a sewage treatment plant. The leachate is usually high in heavy metal ions, chlorides, sulfides, and nitrogen.

Disposal methods aim at removal or reduction of the nuisance and associated health hazard. Methods include:

(1) transportation and dumping; (2) incineration; (3) sanitary landfilling; and (4) salvage and processing.

Historically, much of the handling of trash and garbage has involved hauling it to some location removed from the population center and dumping it. If this dump was on land the waste was often set on fire to reduce the volume of light combustible material. The burning was inefficient and smoky. Wet garbage and metals were left behind to pollute the area and support a heavy population of rats and flies. Waste dumped in water has rotted and polluted streams and ponds. Some waste has been transported to sea and dumped. Some of the lighter and more resistant ingredients have been carried by ocean currents and washed up on beaches.

Incineration is an attractive solution. Burning up the waste to get rid of it suggests finality. But performance raises many problems. Burning such a mixed fuel as municipal trash and garbage produces air pollution that is difficult to control. The ash residue is high in metals and may amount to as much as 40% of the original volume. If the incinerator is over-loaded at any time, the ash will contain organic material that will support rats and flies and putrefies much as does the raw waste. Over-loading during peak hours and poor burning in rainy weather when trash is unusually wet is prevailing practice at municipal incinerators. So incineration is not a full alternative to landfilling but is a method of volume reduction.

Sanitary landfilling buries the waste on land and utilizes soil in the process. Soils vary in their ability to aid in effective disposal. Selection of sites and full use of soil properties are the subject of this paper and will be developed in the following sections.

Municipal waste has many valuable components. Objects and materials are thrown away because they have fulfilled their usefulness to one owner and he has no immediate access to a market. This applies equally to yesterday's newspaper, the carton in which an appliance was shipped, a wornout television set, or fruit and vegetables that spoil in the store before they are sold. Salvage and re-use should be the aim if physical resources are to be conserved. A number of salvage processes are known and have been tried.

Feeding garbage to hogs has been a common practice but is decreasing because of the spread of disease by means of uncooked scraps of meat and because of the practical impossibility of keeping broken glass, bits of metal and indigestible plastics out of the garbage. Sorting newspapers, rags and metals is expensive in labor costs and is usually unprofitable in competition with cheap raw materials that can be processed with a high level of automation. There has been some success with magnetic separation of ferrous metals. It has been hindered by the other metals bonded to the ferrous metal so that extra processing is needed before use.

Composting the organic portion of solid waste to return the nutrients to the soil as a soil conditioner has been tried. Composting plants have had their greatest success in Europe in early post-war years when agriculture was in need and chemical fertilizers were scarce. Some formerly successful plants have closed because of cost of handling the product. At present (1973) all municipal compost plants in the United States are shut down. When they were operating there was some residue that could not be processed and had to be diverted to landfills.

Pyrolysis of the organic portion of solid waste is a possible means of salvaging some of the energy potential in the organic portion of municipal waste. It would produce gas, oil, and coke by destructive distillation. A portion of the waste as collected would not be adaptable to this process and would still have to be disposed of at a landfill or dump.

Some processing does not involve salvage as a main purpose but aims for ease in handling. Grinding municipal waste makes it less bulky. Ground waste packs tight in a landfill, burns better in an incinerator, and is more adaptable to magnetic separation of ferrous metal and ballistic separation of other metals. Baling reduces volume and makes transportation easier and conserves space in landfills.

In spite of the reasons for salvage or incineration, it appears certain that landfill will remain a prime method of disposal for many years. With increasing population and increasing per-capita wastefulness the total need for landfilling will probably increase, even with some progress in the alternatives. So it is well to consider how soil can best serve, and how mistakes or pollution incidents can be avoided.

A sanitary landfill is a waste disposal area which is operated in such a way that odors, smoke, rodents, insect pests, blowing paper, and water pollution, are avoided. Soil is used as the covering and sanitizing material. The solid waste is placed in contact with the soil either in a trench or on an area where some of the soil has been stripped. It is compacted and each day's deposit is covered with at least six inches of soil at the end of the day. Several alternate layers of waste and soil can be built up. Where trenches are used, barriers of undisturbed soil serve as lateral boundaries. When a wider area is filled, soil is placed around the sides and at intervals within the fill so that the waste is confined in masses called cells, which are completely enclosed in soil. When filling of an area is complete, at least 2 feet of soil is placed as final cover. Initial thickness of the cells is usually 6 to 8 feet. As a fill is built some immediate settling results from added weight and from softening of crisp material by moisture. Later settling continues at a slower rate as putrescible materials are destroyed.

Soil has four functions in relation to a landfill.

(1) It serves as container and support.

(2) It provides the most commonly used cover material.

(3) It renovates some of the waste products by direct reaction, by filtration, and by retaining intermediate products providing time and a favorable medium for change.

(4) When it is used as final cover it supports vegetation to re-establish a green landscape with agricultural, forestry, or recreational uses.

Because of their potential for biological activity and complex physical and chemical make-up, soils are more effective in contact with a landfill than strictly geologic materials.

Compared with soil, bedrock or coarse rock fragments present little surface for chemical reaction and have slight filtering effects. Liquids and gases follow fissures in rocks and large voids in gravels or rubble. In soils they disperse through many pores of varying size. When there is occasion to dispose of waste in areas with exposed rock, such as abandoned quarries or strip mines, it is recommended that a soil lining several feet thick be placed on the bottom and against the sides unless the rock is truly impervious and without open joints. When an area to be used for waste is underlain by gravel there is extreme hazard of pollutants being carried down to the groundwater unless a thick layer of fine soil is provided (Hughes et al., 1971). Also there is a hazard of toxic or inflamable gases being carried laterally through gravel and coarse sand strata (Merz, 1954).

Soil to serve as cover needs to have a certain range of characteristics. It needs to be fine enough to make a compact cover that does not readily admit insects and slows but does not stop diffusion of air and water. To permit application of daily cover every day that the landfill is in operation the soil must not be very sticky or it cannot be handled with mechanical equipment during rainy weather. If shrinkage accompanies drying, as in a wet clay, the cover may crack and open the landfill to insects and to the escape of odors. If the soil is loose and non-cohesive, it will be subject to wind and water erosion. All of these considerations lead to a preference for loamy soil, avoiding the extremes of clay, sand, and silt. A moderate proportion of coarse fragments is acceptable, but the diameter of the largest should not exceed 6 inches.

Leachate produced by decomposition of mixed municipal wastes is a solution containing organic acids and alcohols and many dissolved metals. Chloride and sulfate ions are usually the most abundant anions, but phosphate and nitrate are also common. Heavy metal ions vary according to the waste deposited and the stage of decomposition that a landfill has reached. Ferrous iron is a principal metallic constituent of leachate and is usually present in organic complexes that are more soluble than the inorganic salts. Copper, zinc, and lead are other common heavy metals in mixed municipal trash. The decomposition of large quantities of waste in a confined area usually results in anaerobic conditions which are strongly reducing. Much of the nitrogen and sulfur in the fill is converted to ammonia and sulfides. Biochemical decomposition of cellulose, lignin, and carbohydrates usually is only partial, and the acids and alcohols found in the raw leachate require further oxidation to convert them to carbon dioxide and water as the ultimate end products. The average carbon–nitrogen ratio of wastes deposited in a landfill is usually wide. While the fill is in active stages of decay, much of the nitrogen is tied up in the protoplasm of organisms.

The ranges of some pollutants found in raw leachate during two years of operation of a simulated landfill with typical municipal refuse are shown in Table I (Fungaroli, 1971).

During this time the pH varied between 4.5 and 8.4. Flow of leachate varied from very little at the beginning up to the equivalent of water added to the surface in later stages.

Soil underneath the landfill and in the daily cell linings has contact with these varied products and can modify them. Permeable soil that allows air to reach the leachate trapped in pores facilitates oxidation and the shift toward complete elimination of the organic waste. Reaction of organic matter with clay produces relatively stable colloidal complexes.

TABLE I

Occurrence of pollutants (mg/l) in raw leachate from a simulated landfill during 2 years

Pollutant	Occurrence (gm/l)
Total iron	0–1,700
Zinc	0–135
Phosphate	0–130
Sulfate	20–420
Chloride	30–2,300
Sodium	200–3,700
Organic N	20–320
Nickel	0–0.8
Copper	0–7.0
Hardness	150–5,500
Chemical oxygen demand	900–50,000
Suspended solids	10–1,500
Total solids	500–28,000

Heavy metal ions enter into cation exchange with the clay complex. Phosphate ions are tied up with iron in the soil. Light metal ions and chloride, sulfate, and fluoride ions pass through the soil lining with little hindrance.

The quantitative ability of the soil to remove the metal ions from leachate has not been fully evaluated. It is known to be considerable, as leachate collected from leaking landfills where there has been passage through soil is lower in pollutants than the leachate collected directly from waste filled cells where the leachate has not had soil contact. There is need for detailed tests relating the movement of the various constituents of leachate through soils of known cation exchange capacity, degree of base saturation, pH, and clay mineral composition.

The renovating capacity has its limitations as shown by the advance of pollutants behind a moving boundary beneath a landfill (Apgar and Langmuir, 1971).

Observation has shown that many small landfills in good soils have been operated, closed, and then reached a good degree of stabilization without creating a pollution incident. Presumably the soil has been capable of full renovation. Even more landfills have been found to be causing pollution. In most cases some flaw in the soil environment can be identified as a cause, or the quantity of waste products has simply exceeded the capacity of the soil that is available. In these cases there is a need for collection and treatment of the leachate. With large landfills or landfills where soil is not adequate for the renovation needed, it is best to plan for collection and treatment of leachate. When this is the objective, somewhat different soil criteria apply. The aims are to have an impermeable boundary between the base of the landfill and the groundwater and to prevent an excessive volume of water seeping through the fill. If reliance is placed on a tight soil layer or impermeable rock strata it is important to select a site where this natural liner slopes toward a point or relatively narrow zone where the leachate can be collected. Care must be used in construction to avoid breaking the natural barrier. Lateral flow of surface and subsurface water into the site should be intercepted. Such sites are rare, and it is usually necessary to construct a liner of puddled clay, asphalt, rubber, or plastic film.

When leachate has been collected the treatment is simple compared with the job of collection. Treatment with an excess of lime or soda ash to precipitate the heavy metals from alkaline solution and aeration to fully oxidize the organic materials produces an effluent that can be discharged to surface waters. If nitrogen remains high, algae in a stabilization pond can be used to harvest it.

Landfilling should not be considered as a final use for a tract of land but as one stage in changing land use. When a landfill of an area has reached its planned limit, the land should be restored to a productive and esthetically pleasing state. Because some settling will continue after completion, housing is not considered as proper use. Parking areas, athletic fields, playgrounds, golf courses, parks, forests and farmland are suitable uses. All of these are to some degree dependent on the soil used for final cover. In a parking lot the soil may only serve as a base for flexible or semi-rigid pavement. The other uses involve growing vegetation, and the soil needs to meet standards that will provide a favorable environment. A minimum of 2 ft. of soil with moderately high available moisture capacity and moderate fertility is needed. It should have, or be capable of developing, good aggregation and resistance to traffic and erosion. The soil cover should have sufficient permeability to permit rapid diffusion of gases or vents should be provided to drain off hazardous concentrations of methane, carbon monoxide, and carbon dioxide (Merz, 1954).

When all the functions that soil performs in landfill development are considered at once, a very restricted picture of the suitable site results. It portrays a deep well-drained medium textured soil on gentle to moderate slopes. The United States Soil Conservation Service has made interpretations of soil limitations on landfill sites for nearly 20 years. Guidelines were developed to call attention to factors which present hazards and act as limits on use of individual soil mapping units for sanitary landfill. In Pennsylvania the format and principles of these guidelines have been adjusted to embody the standards of the state regulations based on a strict solid waste management act adopted in 1968 (Commonwealth of Pennsylvania, 1968, 1971). Since these are standards based on enforcible regulations they are expressed as suitability ranges. Table II is a summary of the parameters of several soil and site characteristics in three suitability ranges of the Pennsylvania standards. Between the clearly suitable and definitely unsuitable there is a range of limited suitability where use is possible but severely restricted. In Table II the interpretation is for landfills depending on soil renovation. If an impermeable base and collection and treatment of leachate are indicated for a site soil depth, drainage, and depth to seasonal water table are less strict. Blending of soils can make unsuitable soils into satisfactory cover if contrasting materials are available. Steeper slopes can be used if special care is taken for erosion control.

Application of these standards eliminates most soils in humid climates from consideration for sanitary landfill sites with natural renovation. A study made in 1967 (Loughry, 1967) indicates that with application of slightly more liberal standards only 7% of the total area of Pennsylvania has soil physically capable of meeting these requirements. Another

TABLE II

Soil characteristics used in determining site suitability for sanitary landfills in Pennsylvania

Soil site characteristics	Suitable range	Range of limited suitability	Unsuitable range
Depth:			
of developed solum	over 0.9 m	0.4–0.9 m	less than 0.4 m
to hard rock	over 3.7 m	1.2–3.7 m	less than 1.2 m
to fissured rock		over 1.8 m	less than 1.8 m
to gravel or coarse sand		over 1.8 m	less than 1.8 m
Drainage	well drained		all with restricted drainage
Depth to seasonal high water	1.8 m from bottom of planned trench for each 1.8 m lift of refuse	1.8 m minimum	less than 1.8 m
Soil texture	sandy loam, fine sandy loam, loam, silt loam, silty clay loam, sandy clay loam		sand, loamy sand, silt, clay, sandy clay, silty clay, clay loam, fly ash, incinerator residue, organic soils
Slope	0–8%	8–15%	over 15%
Stoniness	nonstony, slightly stony	very stony	extremely stony, stony land
Flooding hazard	none	flooded less frequently than once in 50 years	flooded more frequently than once in 50 years

20% has limited suitability. Existing land use patterns and local objections prevent this use in some of the areas that are physically suitable. Furthermore, the distribution of suitable soil on a regional pattern does not agree with major sources of municipal solid waste.

The soil that falls in the suitable column of Table II on all characteristics is ideal for many uses. There is strong competition for it. It is very good agricultural land or is productive forest land. It is good for building sites; roads are easily constructed on it. It is generally suitable for on-lot sewage disposal. It is adaptable to a wide range of recreational uses.

Detailed soil surveys, as made by the United States Soil Conservation Service, are based on physical and environmental factors that include these items among others. The units mapped can be grouped according to their suitability. These data are available for most of the humid areas of the eastern United States. Several states are completely mapped on a scale of either 1:20,000 or 1:15,840. Pennsylvania has detailed maps on these scales for over 70% of its total area. These maps serve as guides in preliminary screening and evaluation of sites. Applications for new sanitary landfill permits are accompanied by modules giving data on location, geology, groundwater, and more detailed soil descriptions. Joint investigation by a soil scientist and geologist usually is required to verify site suitability before Pennsylvania solid waste management permits are issued for landfills.

Special cases of waste disposal have specific soil relationships. As already mentioned, incineration reduces the volume of waste but eliminates only a fraction of the hazardous ingredients. The ash which is to be disposed of is high in metallic elements. If the burning has been good, there is a considerable reduction of putrescible organic matter. Soil renovation of the leachate from incinerator ash requires capacity for exchange and binding of more heavy metal ions than from an equal volume of mixed trash. Mobilization and transport of the pollutants is earlier and more concentrated than from mixed trash because there is less absorptive paper and wood to hold back water movement.

Metallic sludges from waste water purification processes are obtained by precipitation in presence of excessive alkali. They are almost insoluble at high pH's, but mobilize in an acid environment. If they are to be landfilled without pollution, they should be placed in a soil environment that maintains alkaline conditions and remains free of organic acids. Placement of this kind of waste in an area of basic igneous rocks remote from any water table and with a cover of lime and additional thick cover of soil simulates the environment of natural bodies of ore (Loughry, 1972).

Fly ash from power plants is a very abundant waste in power generating areas. It has some desirable physical properties but usually is contaminated with metal oxides and sometimes with arsenic. In large masses and on steep slopes it lacks stability. It should be filled in areas where the underlying soil and rock are sufficiently well drained that a water table will not rise into the landfill. Final slopes should be covered with soil and have vegetation to prevent erosion.

Agricultural industry wastes from food processing are high in putrescible organic matter and usually very wet when discarded. When landfilled they produce a rapid surge of the intermediate products of decay. If dispersed in a large area of well aerated soil such wastes are readily assimilated. When concentrated in a large mass in a landfill, they exceed the renovating capacity of the available soil and cause pollution incidents.

Animal manures are good soil amendments when incorporated in the surface soil at the rate of a few tons of dry matter per acre. They become disposal problems when large feedlots, suburban dairies, broiler factories, or egg factories concentrate quantities in areas where there is not enough farm land for the traditional use of manure. When manure is landfilled, its nitrogenous content hastens putrefaction with release of obnoxious odors and strong leachate. Its moisture holding capacity keeps it too soft for mechanical placement of earth cover in areas with humid climate.

Treated sewage sludge is sometimes disposed of in landfills. It usually arrives at the landfill in a very wet condition as a soft sludge or slurry. Its nitrogen and phosphorus content promotes rapid breakdown of more carbonaceous wastes. By wetting dryer wastes it hastens the production of leachate. An immediate effect is to make the fill too wet and soft to place earth cover. Thereby it discourages good sanitary landfill practice.

Oily wastes which are difficult to biodegrade can be leached to the groundwater. If the oil is even moderately volatile, fire hazards are increased. A similar hazard in greater degree is involved when waste contains paint thinner, cleaning fluids, industrial solvents, or liquid fuels. Soil renovation of these wastes is usually minimal with volatilization or leaching to the groundwater the usual fate. In some instances lateral diffusion through the soil results in explosive or toxic conditions in basements or buildings.

Even more specialized hazards are involved when pharmaceutical drugs, insecticides, herbicides, polychlorinated biphenols, and radioactive wastes are added to landfills, either by accident or intent. Soil absorbs and degrades some of these, as others pass through unchanged. Handling of radioactive waste is under the control of the Atomic Energy Commission, which requires disposal in containers sufficiently durable to prevent spilling until radioactive decay has lowered activity to safe levels. Meanwhile shielding usually involving thick soil or rock cover is provided. All very hazardous wastes should be barred from landfills except where special handling and treatment are provided to alter them and render them safe.

REFERENCES

Apgar, M.A. and Langmuir, D., 1971. Ground water pollution potential of a landfill above the water table. *Proc. Natl. Ground Water Symp. Ground Water*, 9 (6): 76–96.

Commonwealth of Pennsylvania, 1968 (revised 1972). *Pennsylvania Solid Waste Management Act as Amended. Solid Waste Publication 1.* Commonwealth of Pennsylvania, Department of Environmental Resources, Bureau of Land Protection and Management, Harrisburg, Penn., 17 pp.

Commonwealth of Pennsylvania, 1970. *A Plan for Solid Waste Management in Pennsylvania. Solid Waste Publication 3.* Commonwealth of Pennsylvania, Department of Health, Bureau of Housing and Environmental Control, Harrisburg, Penn., 162 pp.

Commonwealth of Pennsylvania, 1971. *Solid Waste Management, Title 25. Rules and Regulations, Part 1.* Department of Environmental Resources, Subpart C. Protection of Natural Resources, Article 1. Land Resources Chapter 75., pp.75.1–75.16.

Fungaroli, A.A., 1971. *Pollution of Subsurface Water by Sanitary Landfilling.* Research Grant EP-000162, United States Environmental Protection Agency, Washington, pp. 132 + 140.

Hughes, G., Tremblay, J.J., Anger, H. and D'Cruz, J., 1971. *Pollution of Ground Water due to Municipal Dumps.* Tech. Bull., 42. Inland Waters Branch, Department of Energy, Mines and Resources, Ottawa, Ont., 98 pp.

Loughry, F. Glade, 1967. The soil factor in sanitary landfill. *Proc. Penn. Acad. Sci.*, 41: 156–160.

Loughry, F. Glade, 1972. Soil science and geological principles applied to disposal of metallic sludges. *Proc. Tech. Conf. Environmental Conservation, 1st.* The Penn State University Chapter and Keystone Chapter, Soil Conservation Society of America, pp.31–33.

Merz, Robert C., 1954. Report on the investigation of leaching a sanitary landfill. *Calif. State Water Pollut. Control Board Publ.*, 10: 92 pp.

Geoderma, 10 (1973) 141–150

DIKE BREACHES AND SOIL CONDITIONS

P. VAN DER SLUIJS and I. OVAA

Soil Survey Institute, Wageningen (The Netherlands)

(Accepted for publication October 2, 1973)

ABSTRACT

Van der Sluijs, P. and Ovaa, I., 1973. Dike breaches and soil conditions. *Geoderma*, 10: 141–150.

Although soil conditions are certainly not the only factor in the occurrence of dike breaches, the fact still remains that breaches occur more frequently on some soils than on others (Tables I, II and III). The soils concerned are highly permeable soils that permit seepage under the body of the dike. This seepage is considered to be the primary cause of dike breaches attributable to soil conditions and accounts for the higher percentages of breaches on soils with sandy materials at shallow depths.

INTRODUCTION

More than half of The Netherlands is made up of Holocene deposits which are so low-lying that dikes are needed to protect them against flooding by rivers or the sea (Fig.1). Breaches in these dikes have been known throughout Dutch history. In many places the landscape still bears scars that reveal where these catastrophes happened. The inrush of water through a breach in the dike scours a deep hole and when repairs were being made afterwards the new dike would frequently be built around this hole. The original straight dike line is interrupted by an arc-like section around one side of the deep scour hole, called "wiel" in Dutch (Fig.2 and 5).

In the marine area in the southwest of the country, an investigation was made of the nature of the soils at sites where the presence of "wiels" revealed former dike breaches. It was found from a study of soil maps and complementary field work that dikes had broken more frequently on certain types of soils than on others. A cursory look at soil maps of the Dutch fluviatile area led to a similar conclusion.

GENESIS AND CHARACTERISTICS OF SOILS

The genesis and characteristics of the marine and fluviatile soils of The Netherlands have been described in detail by Edelman (1950a).

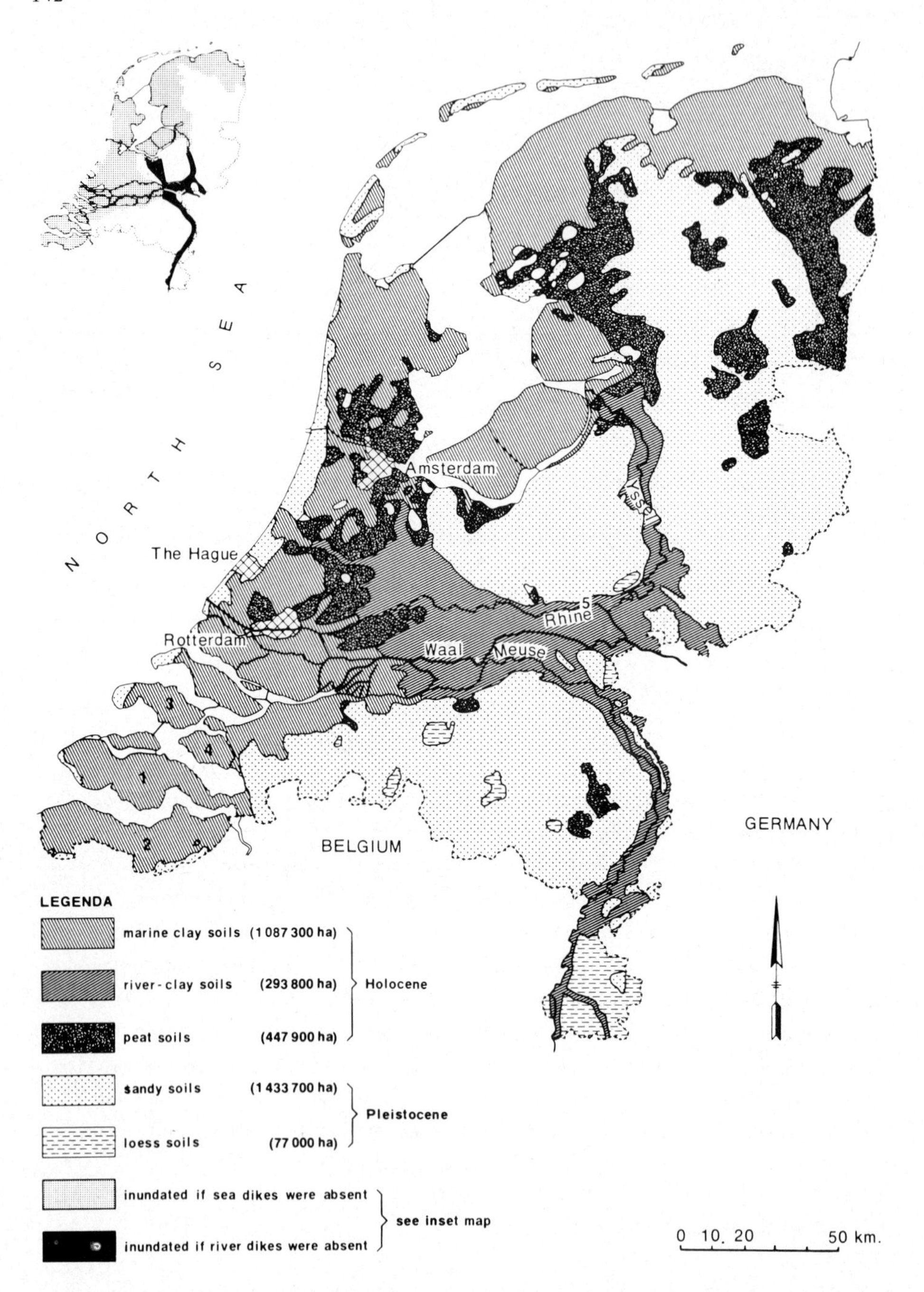

Fig.1. General soil map of The Netherlands. Map top-left: the Holocene area that would be inundated in the absence of dikes (after Van Heesen, 1970). *1* = Zuid-Beveland; *2* = Zeeuwsch-Vlaanderen; *3* = Schouwen-Duiveland; *4* = Tholen.

Fig.2. Polder in eastern part of Zeeuwsch-Vlaanderen. Arc-like sections (*A* and *D*) in the dike indicate former breaches. They are located at an intersection of a dike and a partly silted-up creek (*B* and *C*). Photo: Allied Air Forces. Archives: Soil Survey Institute, The Netherlands.

Marine soils

The marine clay was deposited in a tidal area influenced by the twice-daily rise and fall in sea level. The difference between average high tide and average low tide in the southwest of the country is about 3.5 m. A distinction in land type is made in this landscape between Old Land and New Land.

Old Land

Spread along the coastline at the start of the Sub-Atlanticum was a belt of peat, tens of kilometres wide. Rivers flowed through this peat belt on their way to the sea. The rising sea level brought the peat area within reach of the sea and much of the peat was eroded. On the peat islands that remained, clayey material was deposited by tidal creeks. As these islands were relatively high-lying, they were inundated only during high tides and it was possible for people to settle there. Round about the 10th century the settlers began to build dikes to protect their islands against the progressively higher tides.

Three types of soils are distinguished: creek-ridge soils, pool soils, and transition soils (Fig.3A). Creek-ridge soils were formed in and alongside erosion creeks that gradually silted up and they now exist as ridges in the landscape; their surface layers are sandy loam that merges into lighter subsoils which sometimes consist of sand; their permeability is high, especially in the subsoils. Pool soils are found on the former peat islands; they have a cover of silty clay of poor permeability which changes at depths of between 50 and 150 cm into compressed peat, also poorly permeable. Transition soils are found between the

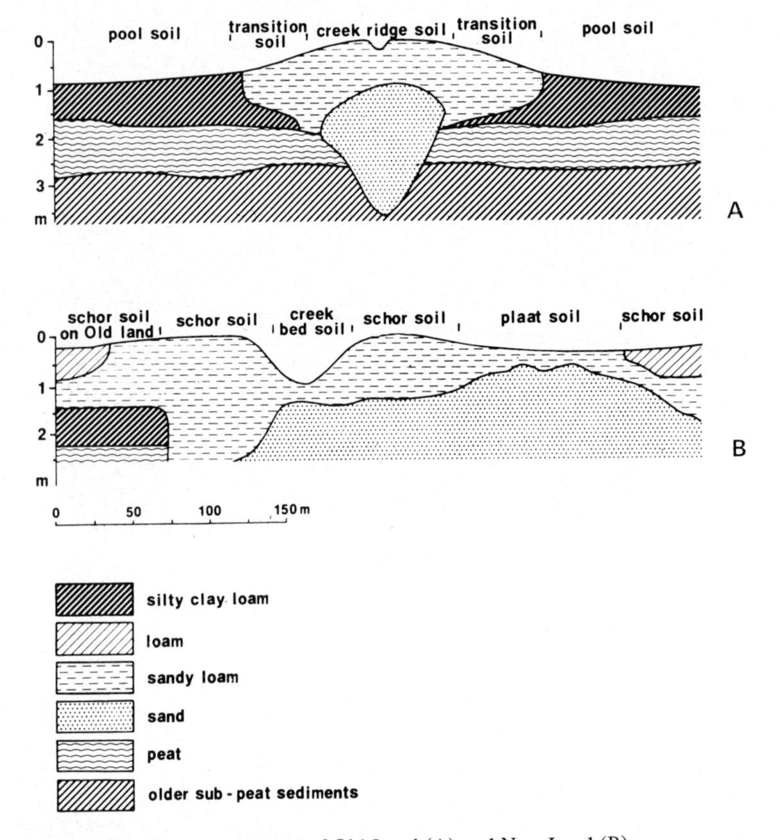

Fig.3. Schematic cross-section of Old Land (A) and New Land (B).

creek-ridge soils and the pool soils; they are loamy soils on a layer of peat, which here occurs at a greater depth than in pool soils; their clay content and permeability are intermediate between creek-ridge soils and pool soils.

New Land

In the gullies between the peat islands – some of which were eroded to a depth of 30 m – sedimentation was also taking place. The powerful tidal currents carried particles of sand, silt, and clay, but because of the strength of the current, only the sand particles could be deposited in the deep gullies. However, as the silting process continued, the periods of inundation grew shorter and the strength of the currents diminished, allowing silt and clay also to be deposited. The level at which the change from sand to clayey material occurred varies greatly (Fig.3B). In places where the current was strongest, thicker beds of sand were deposited than elsewhere.

Because of the high natural fertility of these deposits, each time an area of reasonable size had reached a stage in the silting process that assured natural drainage at low tide, it was immediately enclosed and reclaimed. And so, throughout the centuries a large area of saltings (coastal marshes) has gradually been recovered from the sea. New Land comprises "schor" soils, "plaat" soils and creek-bed soils. "Schor" soils consist of sandy loam, loam, or silty clay loam to a depth of at least 80 cm, the sand content increasing with depth. "Plaat" soils are characterized by a subsoil of sand shallower than 80 cm. Creek-bed soils are the remains of partly silted creeks, dammed while silting was still in progress; their texture profile is the same as the "plaat" soils and they form low-lying areas in the landscape. The presence of the creek-bed soils indicate that the New Land had been unable to complete its sedimentation phase as the Old Land had done before it, former creek beds in the Old Land being completely silted. New Land soils generally are calcareous and have a high permeability.

Fluviatile soils

In The Netherlands the Holocene fluviatile sediments form a layer 3–5 m thick in several broad valleys cut into Pleistocene sands. These are deposits of lower Rhine and Meuse branches which, before being harnessed by artificial levees meandered freely over the low Pleistocene upland. The rivers had built up a network of sandy tracks through back-swamp areas of extremely fine-textured sediments. After completion of all artificial levees in about the 14th century, the course of the meandering rivers was fixed and the silting process only continued in the narrow band of foreland.

The sandy tracks are the higher-lying natural levees of the (former) rivers. Between them are lower-lying deposits of fine-textured soils of back swamps. These so-called river-levee soils and basin-clay soils (Edelman, 1950a) are the fluviatile counterparts of the marine creek-ridge and pool soils distinguished in the Old Land.

TABLE I

Number of dike breaches in the areas Zuid-Beveland and Zeeuwsch-Vlaanderen per 100 km² of soil units.

Landscape and soil unit	Zuid-Beveland			Zeeuwsch-Vlaanderen		
	km²	number of breaches	number of breaches per 100 km²	km²	number of breaches	number of breaches per 100 km²
Old Land:						
Creek-ridge soils	34	16	47	7	0	–
Transition soils	35	2	6	17	0	–
Pool soils	70	1	1	5	0	–
New land:						
"Schor" soils	165	8	5	480	8	2
"Plaat" soils	50	9	18	110	13	12
Creek-bed soils	6	10	167	20	7	35

AVAILABLE DATA

Tables I and II show the number of known dike breaches in several marine-clay areas. Table I relates the breaches to the area occupied by each soil type. Table II gives the relation between the number of breaches per 100 km of dike and the soil type found along the inner side of the wall.

Table III relates the breaches that have occurred in two types of New Land soils ("schor" soils and "plaat" soils) to the depth at which sand is found.

Fig.4 shows that breaches seem to occur frequently at sites where the dikes intersect creek ridges and creek beds. The two readily recognizable breaches depicted in

TABLE II

Number of dike breaches in the areas Schouwen-Duiveland and Tholen, per 100 km dike

Landscape and soil unit	Schouwen-Duiveland			Tholen		
	length of dikes in km	number of breaches	number of breaches per 100 km dike	length of dikes in km	number of breaches	number of breaches per 100 km dike
Old Land:						
Creek-ridge soils	9	5	56	22	4	18
Transition soils	38	4	11	20	4	20
Pool soils	14	0	–	14	0	–
New Land:						
"Schor" soils	54	2	4	72	7	10
"Plaat" soils	32	11	34	32	6	19
Creek-bed soils	5	8	160	6	7	117

TABLE III

Percentile distribution of dike breaches over "schor" soils and "plaat" soils

	"Schor" soils		"Plaat" soils	
	without sand or sand > 120 cm	sand between 80 and 120 cm	sand between 40 and 80 cm	sand < 40 cm
Area (%)	60	15	15	10
Percentage of dike breaches	22	17	27	34
Percentage of dike breaches per 100 km²	5	17	27	51

Fig.2 also occurred at places where the dike intersected a relic tidal creek (points A and D). Its course is still visible in the reclaimed land along the line A B C D. Fig.5 gives a corresponding picture for the river-clay area, one of the breaches being found where the river dike intersects a river ridge, the other where the dike was built on the levee of the present-day river.

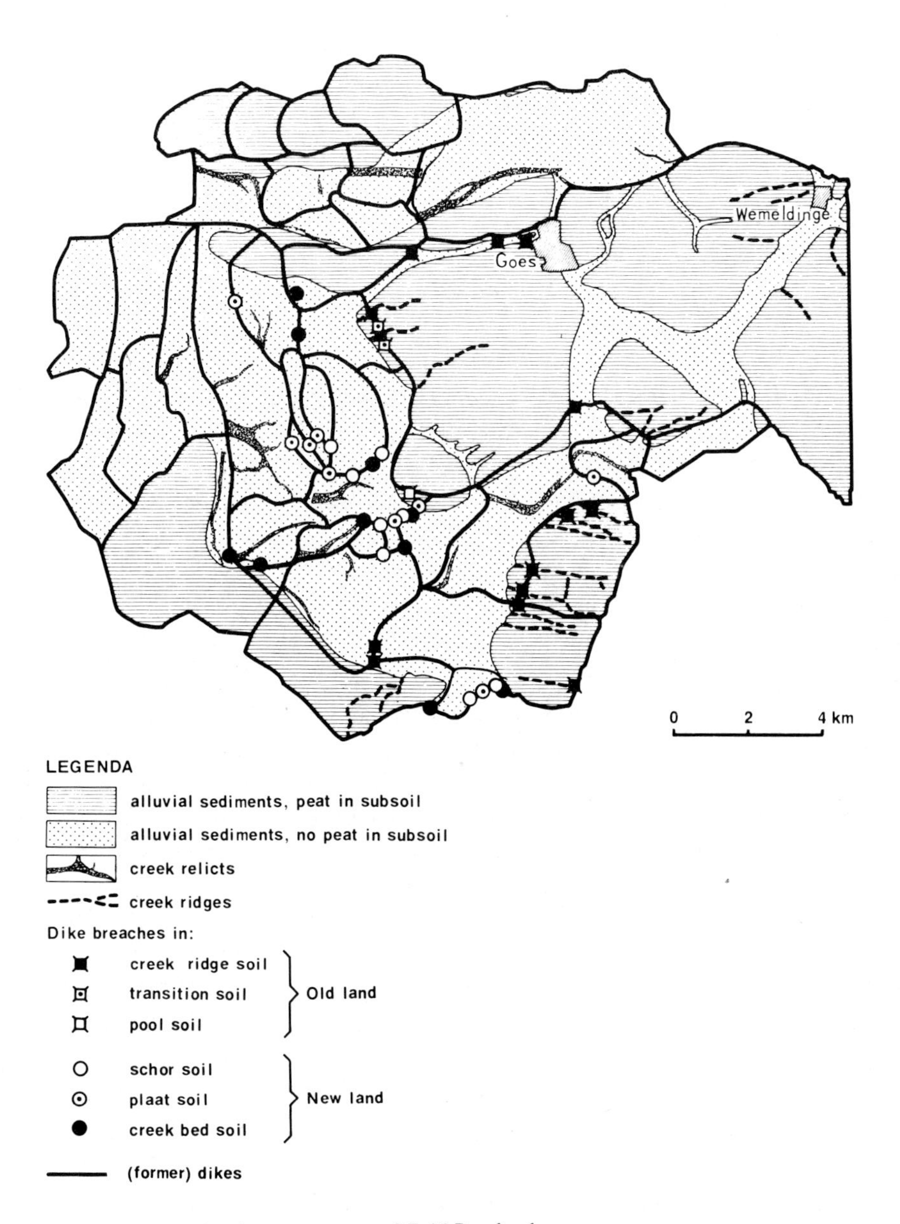

Fig.4. Dike breaches in western part of Zuid-Beveland.

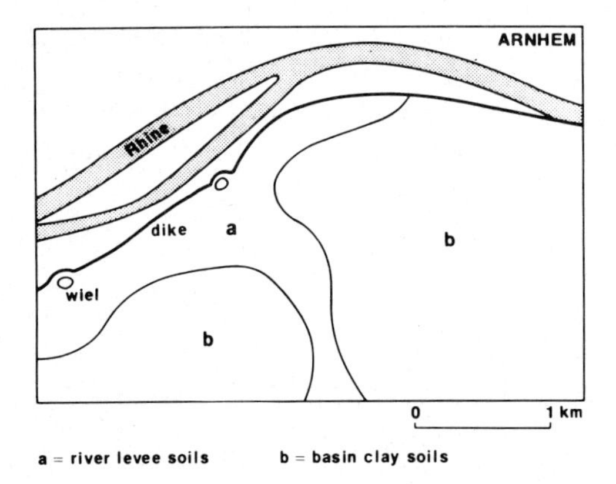

Fig.5. Two breaches of a river dike on a river levee soil.

DISCUSSION

Tables I and II indicate that dike breaches occur more often on certain types of soil than on others, indicating a link between breaches and soil conditions. Other, non-soil, factors naturally have their influence. To mention a few: how the dike is situated with respect to the path of storms; whether saltings are present or not; and how long the dike has served. Even so, the data enable certain conclusions to be drawn. The number of breaches in the various soils shows the same arrangement whether they are related to the area (Table I) or to the length of dike (Table II).

In the Old Land there are only rare instances of dike breaches on pool soils. Breaches are more frequent on transition soils and most frequent of all where the dike has been built on creek-ridge soils.

In the New Land the smallest number of breaches have occurred on the "schor" soils. Considerably more have been noted on "plaat" soils, but the greatest numbers has been recorded on creek-bed soils. The breaches on "schor" and "plaat" soils are correlated positively with the depth to sand (Table III).

The soils on which most breaches seem to occur lie in or near the relic tidal creeks. Creek-ridge soils originated at places where water had eroded the peat subsoil (and often also part of the deeper-lying older deposits), which was then replaced by sand and sandy loam (Fig.3A). These peatless stretches may be broad or narrow. The creek-bed soils (Fig.3B) also originated in former tidal creeks which were not entirely silted at the moment of diking. Some of these sand-filled beds are tens of metres deep. The shallow depth at which sand occurs in "plaat" soils indicates a strong-flowing current of water during the deposition stage - the parent material being deposited as sand banks in the middle of the large creeks, which later silted up. Here too, the sand deposits extend to great depths.

How is the relationship between dike breaches and soil conditions to be explained? Before any explanation can be made, it is first necessary to retrace our steps and check whether our approach to the question is the right one. We must decide whether, by considering only those breaches that have left visible signs in the landscape (scour holes), we have not inadvertently made a biased selection. This might well be true if, because of differences in their resistance to water erosion, some soils are scoured out and others are not. There are undeniably variations in the erosion resistance of soils, yet it is hard to imagine that "schor" soils, for instance, have such enormous resistance to erosion that no traces of dike breaches are to be found in them. So we can disregard the possibility of a lopsided selection.

In the opinion of both Edelman (1950b) and Kuipers (1960), the main cause of dike breaches is a weakening of the dike structure by seepage under the dike. One of the signs indicating seepage is a brown precipitate at the bottom of a water course, caused by the oxidation of iron compounds conveyed in the seepage water. Another sign is no ice forms on the seepage patches during periods of frost or, if present, the ice is friable, has an uncommon colour and may have an hummocky surface. In water courses along dikes, seepage patches are encountered precisely at those places where the water course crosses a sandy strip of a filled-in creek or where sand occurs at a shallow depth.

In the river areas seepage flow is sometimes so strong that on the bottom of the ditch a cone of sand with a crater formed by the outflowing water can be found (boiling of sand) (Van Schaik, 1948).

When the water outside the dike is at a high level and seepage then occurs the flow or hydrodynamic seepage pressure underneath the dike may create "quick sand" in the subsoil. The stability of the dike decreases, fissures appear and soon afterwards the dike may collapse.

Other possible causes seem to be less likely. At the intersection of a dike and a creek bed, some connection might be sought with an extra subsidence by compaction of dike material because the original surface at that spot was lower than its surroundings. If that were so, one could expect to find fewer dike breaches at spots where the dike intersects the ridges of creeks or rivers, because the original surface there was higher than its surroundings. But this is not so. Similarly, there should be fewer breaches on "plaat" soils than on "schor" soils because the subsoil of "plaat" soils (sand) settles less than that of "schor" soils (sandy loam and clay).

Another possible explanation is that the dike construction suffered at those places where there was a lack of suitable building materials in the vicinity. But this does not apply at those sites where the dike intersects narrow creek or river ridges, nor at creek beds. Yet many breaches have occurred at such sites.

Whatever the cause of dike breaches may be, one fact remains, and that is that breaches occur on certain soil types in relatively greater numbers than on others. The soil scientist working in the field, through his knowledge of soil conditions, can make a major contribution towards the recognition of those places where dike breaches are most likely to occur.

REFERENCES

Edelman, C.H., 1950a. *Soils of the Netherlands.* North-Holland, Amsterdam, 177 pp.

Edelman, C.H., 1950b. *A soil survey of the Bommelerwaard boven den Meidijk.* Stichting voor Bodemkartering, Wageningen, 137 pp. (in Dutch with a summary in English).

Kuipers, S.F., 1960. *A Contribution to the Knowledge of the Soils of Schouwen-Duiveland and Tholen According to the Conditions prior to 1953.* Stichting voor Bodemkartering, Wageningen, 192 pp. (in Dutch with a summary in English).

Van Heesen, H.C., 1970. Presentation of the seasonal fluctuation of the water table on soil maps. *Geoderma*, 4: 257–278.

Van Schaik, H.C., 1948. "Sandboil" as the cause of dykeburstings. *Boor Spade*, 1: 164–170 (in Dutch with a summary in English).

Geoderma, 10 (1973) 151–159

INSURANCE CLAIMS FROM EARTHQUAKE DAMAGE IN RELATION TO SOIL PATTERN

R.D. NORTHEY

Soil Bureau, Department of Scientific and Industrial Research, Lower Hutt (New Zealand)

(Accepted for publication August 16, 1973)

ABSTRACT

Northey, R.D., 1973. Insurance claims from earthquake damage in relation to soil pattern. *Geoderma*, 10: 151–159.

An examination of the earthquake damage insurance claims in the City of Wellington, New Zealand, for the period 1940–1970 was made to discover if a local pattern existed which could be correlated with subsoil conditions.

Much of the urban area of Wellington is situated in hilly to steep country with a narrow coastal strip of relatively flat land and a few small valleys. The positions of damaged and undamaged buildings were located on the soil map of the area and the numbers of claims per 1,000 structures were used as a measure of relative severity of earthquake effect between soil parent material mapping units. A very significant correlation was found between the incidence of claims per 1,000 structures and the soil parent material, with severity of effect decreasing with increasing strength and rigidity of soil parent material.

INTRODUCTION

Variations in local and regional soil conditions can exert a marked influence on the patterns of earthquake damage. Analysis of the after-effects of many earthquakes (Duke, 1958; Seed, 1970; Enkeboll, 1971) has shown that the damage patterns of particular types of structure can often be correlated with the density and depth of the underlying soil strata. In association with a seismic microzoning survey (Grant-Taylor, 1973), an examination of earthquake damage insurance claims in the City of Wellington, New Zealand for the period 1940–1970 was made to discover if a local pattern existed which could be correlated with subsoil conditions. Most of the claims were for damage to residential construction so that a reasonable coverage of the principal urban and suburban areas has been possible. Earthquakes of felt intensity Modified Mercalli Scale V or less produced very few claims so that attention has been concentrated on higher felt intensity earthquakes.

The typical New Zealand house has remained of remarkably similar construction since the days of early settlement in mid to late 19th century. It is built largely of wood with galvanised corrugated iron roof. It is wood-framed, with wooden floors and wood external weatherboard sheathing. Wooden piles have now been replaced by concrete, and wooden internal linings by gypsum plaster. Chimneys may be concrete, brick or pumice block now with reinforced shafts, though older buildings have unreinforced chimneys. In this type of construction, with little unreinforced brickwork, earthquake damage is normally relatively slight even in the event of heavy shocks; it usually comprises damaged chimneys, cracked plaster or veneers, cracked drainage pipe connections and occasionally collapsed pile foundations.

Wellington is sited on the southern and western margins of a harbour where easy land for urban development is very restricted by steep hills and drowned valleys. This has forced much of the housing on to hill soils and steepland soils where housing sites frequently require extensive earthworks. Seeking a soil pattern basis for earthquake damage has not been straightforward since most of the early constructed buildings tended to be located on the flatter sites on weaker soils. One would therefore expect more damage on weaker soils, both because of the age of the structure and because of the site. The foundation characteristics of soil sites are influenced by the B, C and D horizons whereas pedological classification of soils emphasises A and B horizons. In this study the C and D horizons are grouped to examine their relationship to earthquake damage. The underlying bedrock, an extremely folded and fractured Mesozoic greywacke, has been deeply weathered, partially eroded and discontinuously covered by Late Pleistocene alluvium, colluvium, loess, and minor amounts of volcanic ash. The steepland soils have formed on slopes generally steeper than 30° directly from weathered greywacke bedrock. The soils on rolling to moderately steep slopes are principally yellow-brown earths and related hill soils derived from the Late Pleistocene drift. The Recent soils on flat and easy rolling land have been derived from Holocene windblown and alluvial sediments containing frequent peat deposits (Taylor and Pohlen, 1968).

NATURE AND ENGINEERING PROPERTIES OF SOIL PARENT MATERIALS

Fill

Only some 7% of the mapped area is naturally relatively flat land and more than 70% is moderately steep to steep. To provide flat land, substantial cutting and filling has been undertaken over the years. At the present time individual fills of more than about 5 ha in area occupy more than 4% of the mapped area. Old fills were constructed with little attempt at formal compaction and occur at a wide range of strength, density and permeability. However, it appears that where older substantial structures are founded in such old fill the loads have normally been carried through to stronger material underneath. Modern structures tend to be founded on the dense Late Pleistocene alluvium, or weathered greywacke where it is close to the ground surface. The bulk of older inland fills are sites of rubbish tips many of which have been very suitably developed as recreation areas.

Extensive compacted fill under engineering supervision is currently being undertaken in several inland areas for housing development. Much of weathered greywacke in Wellington can be compacted with modern earthmoving machinery to rather better densities and strengths than those of the adjoining natural ground. When these fills are properly designed and compacted, they offer no special foundation problem. However, Wellington has not suffered a major earthquake since many of these fills were constructed so their performance under earthquake vibration has not been proved. It is likely that poorly compacted fills, and poorly designed fills on steeper slopes, will give stability problems at such a time, especially if the earthquake has been preceded by abnormally heavy rainfall.

Holocene alluvium

The Holocene alluvium near the surface of valley bottoms has poor engineering properties and is likely to give rise to problems in shallow foundations. This alluvium comprises randomly arranged lenses of soft silts, clays, and peats in coarser alluvium, which itself is loose to compact. Thin gravel layers and lenses frequently occur within a metre of the surface, and many structures are founded in them, even although much weaker materials may occur within the next few metres.

The alluvium which forms terraces now above the present floodplain comprises compact gravels, but there are some weak materials within them. They overlie weathered greywacke or Late Pleistocene alluvium in almost all places.

Late Pleistocene drift

Late Pleistocene drift of alluvium, colluvium and loess covers greywacke bedrock at lower elevations. It can vary in depth from a few metres to several hundred metres. The steepness and shape of the present exposed greywacke topography has been found to be a guide to the steepness and shape of the subsurface greywacke topography. It is thus inferred that many steep-sided small valleys may contain a depth of colluvium comparable with their width.

Though many fine-grained and organic lenses and strata are found in the Late Pleistocene alluvium the predominant superficial texture is coarse-grained. However, much of the original gravel is now so weathered that individual stones cannot be removed from their matrix, and their presence may be indicated in cross-section merely by areas of slightly coarser-textured silty sand within a clay matrix. As a result, normal mechanical analysis procedures show that many of these apparently coarse-grained materials have textures of silt loams to clays. As would be expected such materials show extremely high apparent preconsolidation pressures[*]. Much of the Late Pleistocene alluvium has strength, rigidity and compressibility properties quite comparable with highly and completely weathered greywacke.

[*] When an undisturbed soil is consolidated under a series of increasing static pressures, characteristically compressibility is low up to a certain critical pressure above which compressibility markedly increases. In many sediments this critical pressure is related to the maximum consolidation pressure that the soil has been subjected to and is thus referred to as a preconsolidation pressure. Where the apparent preconsolidation pressure exceeds the present overburden pressure the material is said to be over-consolidated.

Holocene sand

The beach and dune sands vary from 1 to 15 m in depth. They overlie alluvium or, in some places, weathered greywacke. The sands are not very dense, and those in low-lying areas where the water table is high could be expected to compact significantly under load and vibration. Under extreme vibration from a major earthquake some of these deeper sands could be expected to liquefy and lose all stability. However, the frequency of damage claims in the sand area was a little less than average, despite this apparent risk of liquefaction problems. It would seem that during earthquakes not sufficiently intense to produce liquefaction sands behave rather better than fine-grained alluvium, an effect which has previously been noted in Japanese earthquakes by the Tokyo University Earthquake Research Institute (1960, 1971).

Deep loose sands overlying Holocene alluvium present difficult foundation conditions. Shallow sands over Late Pleistocene alluvium or weathered greywacke are less difficult.

Late Pleistocene drift and Mesozoic bedrock

In rolling to moderately steep topography, shallow loess, weathered Late Pleistocene alluvium/colluvium and weathered-in-site greywacke give rise to soils of fairly similar engineering properties. In soil engineering terms (British Standards Institution, 1957), most of these soils are firm to stiff sandy loams to silt loams with low to moderate plasticity. However, profiles developed from deeper relatively coarse-textured sandy loam loess, particularly if it has been deposited on moderately steep slopes, may be prone to local slumping during abnormal rainfall. Thus hill soils formed in such material may be unstable, especially when the natural slope has been locally steepened.

Mesozoic bedrock

Bedrock in the Wellington area, Mesozoic greywacke, has a wide range of depth of weathering and intensity of weathering depending on its original texture and degree of induration. In most places, the uppermost 5–30 m are highly to completely weathered regardless of depth of burial at the present day. The engineering properties of these greywackes vary in accordance with the degree of weathering. In soil engineering terms, highly and completely weathered greywackes are dense to very dense silty angular gravel, and stiff to very stiff sandy loam to silt loam, respectively. The consolidation characteristics of the highly to completely weathered greywacke fit the expected pattern of residual soils that have retained some remnant of the original rock fabric. Apparent preconsolidation pressures are invariably much higher than existing overburden pressures so that the performance of these soils under the load of engineering structures is that of heavily overconsolidated materials.

Slightly to moderately weathered greywacke is the parent material for some hill soils and related steepland soils on sites sufficiently steep for the loess or more completely weathered greywacke to have been stripped. The soils are shallow, and strong materials are found close to the surface. These soils basically present the best foundation conditions but the general steepness of the topography implies extensive earthworks for urban development.

MAJOR EARTHQUAKES IN WELLINGTON, NEW ZEALAND SINCE 1940

1942-earthquakes

A report by the City Engineer (Luke, 1943) records broad details of the damage pattern of the two 1942-earthquakes (both Modified Mercalli Scale VII, Richter Scale 7, distance 100 km) of June 24 and August 2. The principal source of residential damage was to chimneys (some 20,000 required repair), many of which fractured at roof level but remained standing, others in falling severely damaged roof coverings and sometimes the framework of the building. Other residential damage included that to toilet fittings and windows, cracks in plaster and brickwork, and damage from items falling from shelves. It has been estimated (Aked, 1945) that nearly one-third of the total residences in the city were affected.

It was quite clear that there were marked differences in the percentage of damaged buildings between districts and it was felt at the time that this was due to the differences in relative age of buildings in the various districts. Buildings in older districts would be expected to show signs of greater damage especially those built prior to the turn of the century at which time cement was beginning to replace lime in mortars used. The older built-up areas suffered more than the average of 30% damage, while younger areas suffered less than 10% damage.

However, age alone could not explain some of the marked differences between districts. In the group of districts where the majority of residences were built prior to 1903 the range of reported damage was from 2% to 84%, where the majority were built between 1903 and 1920 the range was about 5–63%, and where the majority were built between 1920 and 1931, it was 3–34%. Only among this youngest group of houses, where almost all of the districts suffered less than average damage, did there seem to be a fairly clear correlation with age. This damage pattern showed some marked similarities to the soil parent material pattern. High incidence of damage was confined to the areas from alluvium, wind-blown and beach sand and to some of the rolling hills blanketed with loess and alluvium. Low incidence of damage was confined to the steeper hilly suburbs where weathered bedrock is close to the surface.

In the central city area Johnston (1960) reported on an investigation of some 900 buildings for the Earthquake and War Damage Commission. Some 40% had suffered damage in these earthquakes, and Johnston detected a soil pattern of damage. He showed that buildings in areas of reclamation of the old sea bed suffered more than those located on natural ground.

1943-earthquake

No evidence could be found of significant residential damage arising from the earthquake (Modified Mercalli Scale VI, Richter Scale 5.5, distance 30 km) of February 26, 1943. The bulk of weak chimneys had already been destroyed and any damage observed was still ascribed to the more intense earthquakes a few months earlier.

1966-earthquake

Since 1945 the Earthquake and War Damage Commission has undertaken compulsory earthquake insurance of all private buildings. This systematic recording of claims allows for a more comprehensive approach in damage appraisal. However, the number of claims made as a result of the earthquake of April 23, 1966 (Modified Mercalli Scale VI, Richter Scale 6, distance 50 km) was too small to allow any estimate of the soil pattern. Most of the claims were residential and related to the usual list of damaged chimneys, toilet fittings, hot water cylinders, mirrors and general cracking in plaster and brickwork.

1968-earthquake

About 900 claims on the Earthquake and War Damage Commission were made from the mapped area (Milne, 1973) on account of the earthquake (Modified Mercalli Scale VI, Richter Scale 5.5, distance 30 km) of November 1, 1968.

These claims were enough to justify investigating a soil pattern though insufficient to allow any further analysis into possible patterns related to values, type of claim, or type of structure. In the analysis carried out, only the relative frequency of claims has been investigated. If a claim was made it is likely that the earthquake effect was more significant there than at locations where no claim was made.

Locations of earthquake damage claims were plotted on a transparent overlay on a 1:15,840 cultural detail map. The numbers of claims in each valuation district were then counted for comparison with the numbers and age distribution of buildings. In 1968 there were a total of 25,000 houses in the mapped area, very little different to the number of houses in 1942 due to the northward migration of population outside the mapped area. Some 90% of the total number of houses in the mapped area were accounted for in this analysis. The overall damage pattern was remarkably similar to that of the 1942-earthquakes. Attempts were made to assess the importance of the age factor in this pattern. Various parameters were selected to characterise the age distribution of houses and then the proportion of damage claims was recalculated for groups of districts showing similar age distributions. From these comparisons, it appeared that there was a factor of 2 to 3 for this earthquake between the relative average claim density in older (pre-1914) houses and districts, and that in younger (post-1950) houses and districts, in similar topography.

The transparent overlay of damage claims and cultural detail was then compared with the soil map of Wellington prepared by Milne (1973). Soil maps themselves generally

TABLE I

Incidence of insurance claims, 1968-earthquake, Wellington, N.Z.

Soil parent material	Soil mapping unit* (mainly soil associations)	Area (%) (total 7,768 ha)	Residences (%) (total 22,200)	Claims (%) (total 895)	Average value of claims ($N.Z.) (total $96,920)	Claims per 1,000 structures
Fill	fill	4.1	3.8	10.7	225	116
Holocene alluvium	Waiwhetu silt loam	3.2	9.2	15.0	110	66
	Waiwhetu silt loam – Gollans silt loam	0.2	0.4	2.4	120	263
	Waiwhetu silt loam – Waikanae silt loam	0.7	2.0	3.3	180	67
	Waikanae silt loam – Waiwhetu silt loam	0.6	1.2	1.8	50	62
	Waiwhetu silt loam – Korokoro hill soils	0.3	0.3	0.6	100	86
	Heretaunga silt loam – Waiwhetu silt loam	0.4	0.3	0.2	75	25
		5.4	13.4	23.3	115	70
Late Pleistocene drift (alluvium, colluvium, loess)	Paremata silt loam	6.0	19.8	28.0	80	57
	Porirua fine sandy loam	1.3	3.6	5.3	90	59
	Paremata silt loam – Porirua sandy loam	0.5	1.7	0.3	70	8
	Paremata silt loam – Korokoro hill soils	1.5	3.6	3.9	60	44
	Korokoro hill soils – Paremata silt loam	0.3	0.9	0.9	45	42
	Paremata hill soils – Korokoro hill soils	0.4	0.3	–	–	–
	Ngaio silt loam	1.6	2.6	2.4	60	37
	Ngaio silt loam – Korokoro hill soils	2.2	5.1	3.1	40	25
	Ngaio hill soils – Korokoro hill soils	1.6	2.2	0.5	45	10
		15.4	39.7	44.4	74	45
Holocene sand	Waitarere soils – Hokio soils	3.6	9.7	8.3	70	34
Late Pleistocene drift and Mesozoic bedrock	Korokoro hill soils – Makara hill soils	2.1	1.4	0.6	50	19
	Makara hill soils – Korokoro hill soils	24.4	13.5	6.5	140	19
	Porirua hill soils – Terawhiti hill soils	1.0	0.4	–	–	–
	Terawhiti hill soils – Porirua hill soils	3.3	1.1	0.8	80	35
	Terawhiti hill soils – Korokoro hill soils	6.5	11.8	3.6	80	12
		37.3	28.2	11.5	110	16
Mesozoic bedrock	Makara steepland soils	24.4	2.9	1.1	40	15
	Terawhiti steepland soils	8.6	2.3	0.7	50	12
		33.0	5.2	1.8	45	14

*Milne, 1973

reflect the nature of only the upper 1–2 m of material beneath ground surface. However, earthquake response is controlled by the nature, thickness, and distribution of deeper material which overlies bedrock. In addition, information is needed on the depth and intensity of weathering of the bedrock. In Wellington, however, there is a fairly well defined relationship between the nature of materials in the upper 1–2 m and the general nature of deeper materials. Accordingly, the soils on the soil map were grouped into 6 classes based on soil parent material. It was found (Table I) that a good relationship existed between the six classes and the density of earthquake damage claims.

This can be summarised in terms of average claim density (claims per 1,000 structures) as follows:

Fill	three times average
Holocene alluvium	twice average
Late Pleistocene drift	slightly above average
Holocene sand	slightly below average
Late Pleistocene drift and Mesozoic bedrock	less than one half average
Mesozoic bedrock	one third average

There is a factor of at least 8 between the high incidence on fill (116) and the low incidence on weathered bedrock (14). If we compare this with the expected factor of 2 to 3 due to age distribution of houses, it is clear that even if there had been a perfect match of the oldest most vulnerable structures on the worst soils, a factor of 3 to 4 is ascribable to differences between soils.

CONCLUSION

This investigation has shown that the incidence of damage claims from recent earthquakes in Wellington, New Zealand, shows a significant correlation with the pattern of soil parent materials. While knowledge of the nature, thickness, distribution, and intensity of weathering of soil materials over bedrock is generally required to predict earthquake response of a site, in this case the pedological soil map of the area has been shown to be a reasonable expression of the pattern of much deeper soil materials. As a result of this study it is possible to place numerical values, even though approximate, on the relative risk of earthquake damage to residential construction in Wellington according to the underlying soil conditions.

REFERENCES

Aked, W.E., 1945. Damage to buildings in the city of Wellington by earthquake, 1942. *Proc. N.Z. Inst. Eng.*, 32: 132–183.

British Standards Institution, 1957. *Site Investigations*. British Standard Code of Practice, CP2001.

Duke, C.M., 1958. Effects of ground on the destructiveness of large earthquakes. *J. Soil Mech. Found. Div. Am. Soc. Civil Eng., 84 (SM3). Proc. Pap.*, 1730: 23 pp.

Enkeboll, W., 1971. Soil behaviour and related effects in the Peru earthquake of May 31, 1971. *Bull. Seismol. Soc. Am.*, 61: 579–590.

Grant-Taylor, T.L. (Editor), 1973. Microzoning for earthquake effects in Wellington, N.Z. *N.Z. Dep. Sci. Ind. Res. Bull.*, 213: 57 pp.

Johnston, J.A.R., 1960. A brief history of damaging earthquakes in Wellington City and developments in multi-storey building construction in New Zealand. *Proc. World Conf. Earthquake Eng., 2nd, 1960*, 1: 457–471.

Luke, K.E., 1943. Report on damage by earthquakes 1942 in the City of Wellington, N.Z. *City Engineer's Report to Wellington City Council* (unpublished).

Milne, J.D.G., 1973. Soil map of Wellington urban area, New Zealand. Scale 1:15,840 Map 3. In: T.L. Grant-Taylor (Editor), Microzoning for earthquake effects in Wellington, N.Z. *N.Z. Dep. Sci. Ind. Res. Bull.*, 213: 57 pp.

Seed, H.B., 1970. The influence of local soil conditions on earthquake damage. *Proc. Int. Conf. Soil Mech. Found. Eng., 7th, Mexico City, 1969, Speciality Sess. 2, Soil Dyn.*

Taylor, N.H. and Pohlen, I.J., 1968. Classification of New Zealand soils. In: Soils of New Zealand, I. *N.Z. Soil Bur. Bull.*, 26(1): 15–46.

Tokyo University Earthquake Research Institute, 1960. Earthquake damage and subsoil conditions as observed in certain districts of Japan. *Proc. World Conf. Earthquake Eng., 2nd, 1960*, 1: 311–326.

Tokyo University Earthquake Research Institute, 1971. Observation of earthquake motions on various kinds of grounds in Hachinoe City. *Proc. Japan Earthquake Eng. Promotion Soc., Tokyo*, pp.28–37.

Geoderma, 10 (1973) 161–168

TEPHRA MARKER BEDS IN THE SOIL AND THEIR APPLICATION IN RELATED SCIENCES

W.A. PULLAR

Soil Bureau, Department of Scientific and Industrial Research, Rotorua (New Zealand)

(Accepted for publication July 6, 1973)

ABSTRACT

Pullar, W.A., 1973. Tephra marker beds in the soil and their application in related sciences. *Geoderma*, 10: 161–168.

Tephra marker beds in the soil have been used to examine aspects of natural history including the volume and rate of alluvial infilling of basins, age of ground surfaces, progradation of shorelines, time of earth movements, fan-building, and history of vegetation in peat swamps. Their application to archaeology, flood problems and volcanic risk is also briefly mentioned.

INTRODUCTION

In the course of detailed soil surveys (1:15,840) of Gisborne Plains (Pullar, 1962), Wairoa valley (Pullar and Ayson, 1965), and Rangitaiki Plains (Pullar, in press) (Fig.1) thin volcanic ash beds (tephra) were often noted in the soil profile. These beds were later identified and radiocarbon-dated (Vucetich and Pullar, 1964) and thus became useful as marker beds in estimating for example the amount and rate of infilling of basins, the incidence, degree and time of fan-building, and the long and short-term rate of progradation of shorelines. Although tephra marker beds may not be strictly part of the soil in the sense that they do not always take part in soil processes, nevertheless their usefulness as marker beds in providing the detail necessary in quantitative work could only be obtained during the course of soil surveys.

The marker beds shown in Table I, identified in the soil to a depth of about 1 m, have proved to be useful.

TABLE I

Tephra marker beds and ages

Marker bed	^{14}C age years BP (1950): ^{14}C number
Tarawera Ash and Lapilli Rotomahana Mud	(historical, AD 1886)
Kaharoa Ash	930 ± 70: NZ10
Taupo Pumice	1840 ± 50: NZ1548
Waimihia Formation	3440 ± 70: NZ2
Whakatane Ash	4680 ± 100: NZ1358

INFILLING OF BASINS

Infilling of basins with alluvium is a measure of the incidence and degree of erosion in the associated uplands.

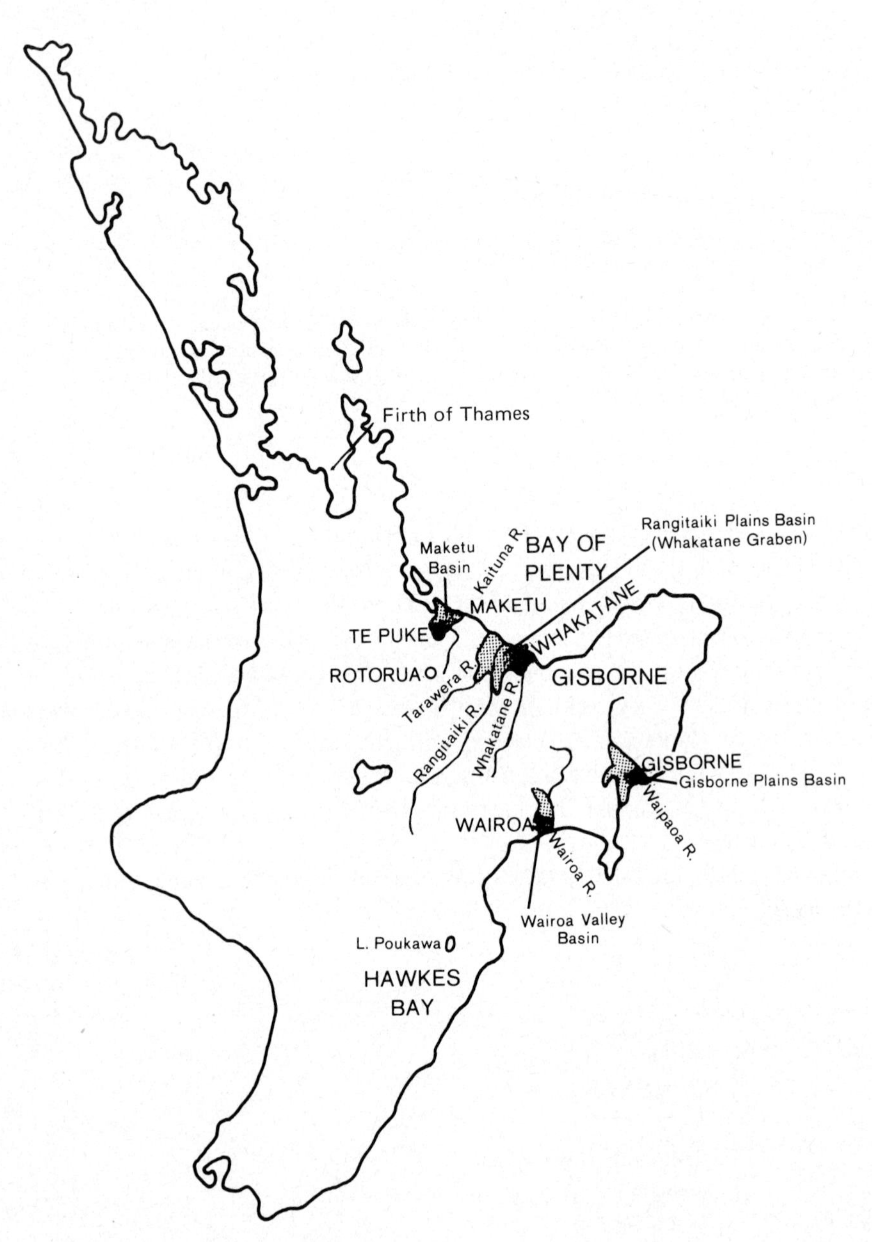

Fig.1. Locality map showing basins.

Gisborne Plains basin (Pullar and Penhale, 1970) (Fig.1)

This basin is associated with the Waipaoa river catchment which was deforested in the late 1800's for pastoral farming. Rocks in the upland areas are sedimentary and of Cretaceous and Tertiary age, those of the former age being highly deformed and noted for accelerated erosion. The basin was examined in great detail using both buried paleosols and tephra formations as marker beds. The age of the buried paleosols was obtained indirectly from dendrochronology. The distribution of the alluvium deposited during stated intervals was displayed on maps in the form of thickness steps ranging from 8 to 30 cm. The volumes and rate of alluvial infilling are given in Table II.

The results show that for the period 1932–1950 the rate of infilling is 5 to 10 times that for any other period. Infilling during the period AD 1650–1820 is attributed to catastrophic storm damage ca. AD 1650 and that during the period 1400 BC to AD 131 is attributed to earth movements.

TABLE II

Volumes and rates of alluvial infilling of the Gisborne Plains basin
(After Pullar and Penhale, 1970)

Period		Length of period (yrs.)	Volume of infilling ($m^3 \cdot 10^6$)	Rate of Infilling		Reliability of estimate
				volume ($m^3 \cdot 10^6$)/yr.	thickness (mm/yr.)	
1	*1480 B.C.–**A.D. 131	1611	342	0.21	1.5	low
2	A.D. 131–1650	1519	159	0.10	0.8	medium
3	1650–1820	170	30	0.17	2.3	medium
4	1820–1932	112	14	0.13	3.7	high
5	1932–1950	18	20	1.14	16.0	high

*Waimihia Formation.
**Taupo Pumice.

Wairoa valley basin (Pullar and Ayson, 1965)

In this basin the Taupo Pumice marker bed was employed to measure the thickness of infilling of pumice alluvium after the Taupo Pumice eruptions of ca. 1800 BP. The most common thickness ranges from 60 to 120 cm with small areas from 1.8 to 3.0 m.

In other small basins in the Gisborne and Hawkes Bay districts the best marker bed is the Taupo Pumice. Infilling is much less marked than on Gisborne Plains and ranges from 30 to 90 cm in swamps and a little over a metre on natural levees of rivers and streams.

Rangitaiki Plains basin (Pullar, in press)

This basin has also been examined in great detail but only a portion of the work has so far been published. The plains are traversed by the Whakatane, Rangitaiki and Tarawera rivers; the first-named flows in a meander trough and the others have built up fans into which they have now degraded.

The Whakatane river drains Jurassic greywacke ranges partially mantled with tephra; the Rangitaiki river has part of its headwaters in greywacke rocks and part in volcanic ignimbrite rocks thickly mantled with tephra; the Tarawera river is lake fed and its tributaries drain thick tephra deposits resting on rhyolite and ignimbrite.

The most useful tephra marker beds on the plains include Tarawera Ash, Kaharoa Ash, and Taupo Pumice. The floor of the meander trough of Whakatane River near Whakatane has been raised 30 to 100 cm above the Tarawera Ash marker bed in 75 years (Pullar, 1963) and flood water storage in the trough has thus been reduced by about 10%. After the Taupo Pumice eruptions, the Rangitaiki river built up a fan of pumice gravels as thick as 10 m. Little fan building occurred after the Kaharoa eruption. In the Tarawera fan the Kaharoa Ash marker bed has seldom been seen so that the fan must be at least 3 m thick. In back-swamp lowlands and natural levees the most useful marker bed is the Kaharoa Ash which occurs within a metre of the surface.

Maketu basin

Part of the Maketu basin near Te Puke has been examined in detail but the work is not yet published. In general, the Kaharoa Ash marker bed is buried from 15 to 30 cm and the Taupo Pumice from 30 to 150 cm from the surface.

The principal river, the Kaituna, drains an ignimbrite plateau deeply dissected by narrow gorges and mantled with older Holocene and Late Pleistocene tephra (Vucetich and Pullar, 1969).

COMPARISON OF BASIN INFILLING

The Wairoa valley and Rangitaiki Plains basins are filled with a high volume of pumice alluvium derived from the products of volcanic eruptions and most likely deposited shortly after the eruptions. The Gisborne Plains basin has a high volume of alluvium derived from sedimentary rocks and infilling seems to have proceeded in spurts. The Maketu basin, on the other hand, has a relatively much smaller volume of alluvium derived from both tephra and ignimbrite. It would appear that the older, weathered Late Pleistocene tephra is not so erodible as the younger tephra and clings better to the ignimbrite rock.

The most important period for infilling studies is within the last hundred years. In the Gisborne district the uplands were deforested by European settlers and this act induced accelerated erosion; but no-large scale deforestation took place in the Whakatane river catchment. Erosion here is attributed more to frequency of high-intensity rains that are known to sweep over the catchment. Uplands associated with the Maketu basin were only partially deforested and erosion appears to be negligible compared with that in uplands associated with the Whakatane river catchment and Gisborne Plains basin.

AGE OF GROUND SURFACES

On Gisborne Plains and in Whakatane river valley, soils with colour (B) horizons occur on low terraces associated with former flood plains. They are considered to have been beyond flood-reach for about 500 years (Pullar, 1967a; Pullar et al., 1967). They also have the Kaharoa Ash marker bed in the profile and within a metre of the surface. The ground-surface has been named the post-Kaharoa surface and it is suggested that soils elsewhere with similar profiles but without the Kaharoa Ash marker bed could be of the same age.

PROGRADATION OF SHORELINES

On Rangitaiki and Gisborne Plains both coastal and inland dunes are mantled with tephra marker beds. The position where these beds cut out has been suggested as a former shoreline (Pullar and Selby, 1971). On Rangitaiki Plains the dune system was formed over a period of 8000 years and the coast has prograded about 10 km in that time. The rate of progradation over the last 1800 years ranges from 0.5 to 0.7 m/yr and this compares with 0.5 to 0.8 m/yr at Gisborne (Pullar and Penhale, 1970). From 1886 to 1962, the rate of progradation of Rangitaiki Plains is about 1.3 m/yr and that of Gisborne Plains about 1.4 m/yr.

Thus the rate of progradation over the last 80 years appears to be twice as high as the average over the last 1800 years.

EARTH MOVEMENTS

At Gisborne, the Waimihia marker bed and associated layers of organic mud were seen to be tilted at about 3°; no tilting was noted with the Taupo Pumice marker bed (Pullar, (1967b). On hills adjacent to Gisborne, gully-in-gully forms are common with peat infilling in the lower gully. The Waimihia marker bed was noted in the peat. Rejuvenation of gullies suggests that uplift was intermittent from before the Waimihia eruption until the time of the Taupo Pumice eruption.

The Rangitaiki Plains basin is part of the Whakatane graben. Inland dunes mantled with Whakatane Ash and peat are buried to a depth of about 3 m below present sea level, and this position with respect to the sea suggests either a low sea level at the time of dune formation or a sinking of the land by earth movements, or a combination of both. According to Schofield (1960), the sea level in the Firth of Thames was 2.1 m higher 4000 years ago and if this represents a eustatic rise only the Rangitaiki Plains have sunk by about 5 m and the dune system become drowned.

Time of drowning might be gauged by the start of peat formation which has been dated on a buried dune ridge at ca. 3200 BP.

In Europe, the maximum temperature rise in the Hypsithermal interval occurred about 6000 BP but the effects of a sea-level rise resulting in an emergent dune system are not reflected on the Rangitaiki Plains. The problem may be resolved by noting the depths of shells and stones associated with former inter-tidal strands, and on this field work is proceeding.

FAN-BUILDING

Fan-building refers to small fans bordering the hills around the Rangitaiki Plains. The deposits of colluvium contain the Kaharoa Ash and Tarawera Ash marker beds and their positions relative to the surface indicate the degree of fan-building activity. If they are both at the surface then the fan has been inactive for 900 years; if colluvium occurs above the Tarawera Ash then the fan is still active.

HISTORY OF VEGETATION IN PEAT SWAMPS

Plant remains between marker beds in peat swamps in Rangitaiki Plains and Maketu basins have been examined by Campbell et al. (1973). The peat deposits are underlain by the Whakatane Ash marker bed and so are assumed to be no older than 4000 years. The peat is mainly low-moor, sedge peat formed from *Baumea*. In wetter periods *Restionaceae* was established and in drier periods *Leptospermum* and *Gleichenia*. The presence of charcoal fragments between the marker beds indicates that fires have been frequent over the past 4000 years.

IDENTIFICATION OF TEPHRA FORMATIONS IN PEAT DEPOSITS

In areas distant from volcanic eruptive sources it is impossible to identify tephra formations on terraces and hills because the formations are shrouded in the surface soil by organic matter. Formations such as Kaharoa Ash and Taupo Pumice have been successfully identified in peat deposits where their white colours contrast distinctly with the black-coloured, containing layers of peat. Ages of tephra formations have also been determined by ^{14}C dating of the peat bracketing the formation above and below (Vucetich and Pullar, 1971; Pullar and Heine, 1971).

ARCHAEOLOGY

In the soil the evidence for Maori occupation is mainly a deep black greasy topsoil containing burnt stones, shells, and sometimes artefacts. In coastal Bay of Plenty, occupation sites are often covered with Tarawera Ash (Pullar et al., 1967) but none have been found mantled with Kaharoa Ash. On the coast just south of Gisborne no occupation layers were seen below the Taupo Pumice marker bed. Near Lake Poukawa in Hawkes Bay, Price (1963, 1965) has uncovered artefacts and moa bones under the Taupo Pumice marker bed and items related to man under the Waimihia marker bed (Pullar, 1970).

FLOOD PROBLEMS

On Gisborne Plains and in the Whakatane river valley, the post-Kaharoa surfaces are now receiving sediment from the larger present-day floods after having been flood-free for 500 years (Pullar, 1967a,b). It is on these surfaces that towns have been built, and at the time justifiably so, because of the assumption that they had been flood-free for a long time. To keep these towns flood-free now requires expensive flood control measures and means another tax on town dwellers. If possible, other land forms such as terraces contiguous to flood plains, easy rolling hills and coastal dunes, should be considered in planning the expansion of a town.

VOLCANIC RISK

A small eruption, the historical Tarawera eruption, is known to have deposited ash and lapilli to a depth of 15 cm on at least half the area of Rangitaiki Plains (32,000 ha) and to have spoilt pastures near Te Puke with a mantle of Rotomahana Mud 3 cm thick. The Kaharoa eruption would have covered the Rangitaiki Plains with a mantle of ash at least 15 cm thick. After the Kaharoa eruption, the Tarawera river transported a large quantity of sand and pumice gravel to the lower reaches near the coast where swamps have been filled in to a depth of about a metre. On the flood plain at that time braided channels were a feature (Pullar, 1967b).

Although the direct effect of the Taupo Pumice eruptions on the Bay of Plenty and Gisborne districts would have been no more than a light mantling of ash and lapilli, their effect on the regimes of Whakatane and Rangitaiki Rivers was tremendous. For the former river it is estimated that a slurry about 1 m thick was transported to Whakatane and this thickness of fine material would no doubt have killed trees on the flood plain. In the latter river, a fan of pumice gravels 10 m thick was built up where the river debouched on to the Rangitaiki Plains.

The effects of the Whakatane Ash eruption would be very serious to agriculture on Rangitaiki Plains and to Whakatane town if the eruption occurred today. The area would be smothered in ash and lapilli to a depth of 60 cm, and the weight of these materials on the roofs of houses would amount to about 20 tonnes.

CONCLUSIONS

Non-agricultural uses of soil survey are best derived from detailed surveys. These take a long time to execute but the "spin-off" from them is great. In New Zealand, the pedologist lives in the area being surveyed and in so doing gains the confidence of farmers and officers of government and local body agencies from whom much information is obtained. The information gained may be known already to local people but its rationalization by means of detailed soil surveys is valuable to others from afar and certainly to people of the next generation.

REFERENCES

Campbell, E.O., Heine, J.C. and Pullar, W.A., 1973. Plant identification and pollens from peat deposits in Rangitaiki Plains and Maketu basins. *N.Z. J. Botan.*, 12: 317–330.
Price, T.R., 1963. Moa remains at Poukawa, Hawkes Bay. *N.Z. Archaeol. Assoc.*, 6: 169–174.
Price, T.R., 1965. Excavations at Poukawa, Hawkes Bay, New Zealand. *N.Z. Archaeol. Assoc.*, 8: 8–11.
Pullar, W.A., 1962. Soils and agriculture of Gisborne Plains. *N.Z. Soil Bur. Bull.*, 20.
Pullar, W.A., 1963. Flood risk at Whakatane. *J. Hydrol. (N.Z.)*, 2: 47–52.
Pullar, W.A., 1967a. Flood problems. *N.Z. Geographer*, 23: 70–72.
Pullar, W.A., 1967b. Uses of volcanic ash beds in geomorphology. *Earth Sci. J.*, 1: 164–177.
Pullar, W.A., 1970. Pumice ash beds and peat deposits of archaeological significance near Lake Poukawa, Hawkes Bay. *N.Z. J. Sci.*, 13: 687–705.
Pullar, W.A., in press. Soil map and extended legend of Whakatane Borough and environs, Bay of Plenty, New Zealand. *N.Z. Soil Surv. Rep.*,8.
Pullar, W.A. and Ayson, E.C., 1965. Soils and agriculture of Wairoa Valley, Hawkes Bay. *N.Z. Soil Bur. Rep.*, 2(1965): 35 pp.
Pullar, W.A. and Heine, J.C., 1971. Ages, inferred from ^{14}C dates, of some tephra and other deposits from Rotorua, Taupo, Bay of Plenty, Gisborne, and Hawke's Bay districts. *Proc. Radiocarbon Users Conf. Wellington*, pp.119–138.
Pullar, W.A. and Penhale, H.R., 1970. Periods of recent infilling of the Gisborne Plains basin; associated marker beds and changes in the shoreline. *N.Z. J. Sci.*, 13: 410–434.
Pullar, W.A. and Selby, M.J., 1971. Coastal progradation of Rangitaiki Plains, New Zealand. *N.Z. J. Sci.*, 14: 419–434.
Pullar, W.A., Moore, K.W. and Scott, A.S., 1967. Field archaeology in the Bay of Plenty. *Historical Rev.* (Journal of the Whakatane and District Historical Society), XV: 105–114.
Pullar, W.A., Pain, C.F. and Johns, R.J., 1967. Chronology of terraces, flood plains, fans and dunes in the lower Whakatane River Valley. *Proc. N.Z. Geogr. Conf.*, 5th, pp.175–180.
Schofield, J.C., 1960. Sea level fluctuations during the last 4000 years as recorded by a chenier plain, Firth of Thames, New Zealand. *N.Z. J. Geol. Geophys.*, 3: 467–485.
Vucetich, C.G. and Pullar, W.A., 1964. Stratigraphy and chronology of Late Quaternary volcanic ash in Taupo, Rotorua and Gisborne districts, Part 2. *N.Z. Geol. Surv. Bull.*, 73: 43–88.
Vucetich, C.G. and Pullar, W.A., 1969. Stratigraphy and chronology of Late Pleistocene volcanic ash beds in central North Island, New Zealand. *N.Z. J. Geol. Geophys.*, 12: 784–837.
Vucetich, C.G. and Pullar, W.A., 1971. Radiocarbon chronology of Late Quaternary rhyolite tephra deposits. *Proc. Radiocarbon Users Conf. Wellington*, pp.112–118.

Geoderma, 10 (1973) 169–178

THE VALUE OF SOIL SURVEY FOR ARCHAEOLOGY

L.W. DEKKER and M.D. DE WEERD

Soil Survey Institute, Wageningen (The Netherlands)
Institute for Pre- and Protohistory, University of Amsterdam, Amsterdam, and Westfrisian Museum, Hoorn (The Netherlands)

(Accepted for publication October 2, 1973)

ABSTRACT

Dekker, L.W. and De Weerd, M.D., 1973. The value of soil survey for archaeology. *Geoderma*, 10:169–178.

The information gained about the origin and stratigraphy of sediments from a soil survey is extremely useful for archaeologists who study human occupation through the course of time. Soil maps often form a valuable basis for recording or mapping archaeological finds. The information is also helpful in the search for clues to past habitation in deposits now below the surface because of burial by more recent sedimentation. The value of a soil survey for archaeology is illustrated by the results of a study in Midden-Westfriesland, an area in the western part of The Netherlands.

INTRODUCTION

Soil surveyors come across archaeological finds at regular intervals during their field work. Old settlement soils are often recognized by thick black humose layers and greenish yellow ironphosphate stains in the subsoil. Sometimes potsherds are found at the surface or brought up from the deeper sediments.

It is of great importance to the archaeologist that artefacts and settlement traces be plotted on maps. Distribution maps are a main source of archaeological knowledge; they reveal the spatial patterning of material culture.

Archaeological finds as a rule link up with soil conditions. For example, in the marine-clay area in the western part of The Netherlands traces of prehistoric occupation have been found mostly in soils of beach and creek ridges. In the river deposits in the central part of The Netherlands, settlement traces are mainly found in soils and sediments of natural levees along former stream channels and in Pleistocene sandy outcrops. Arable farming was possible on these soils as they have a relatively deep watertable, and they are easy to work because of their texture and structure. Soil surveys have established the geographical distribution of these soils.

Suitability for human habitation occasionally changed in the course of time when areas were flooded by sea or river water. Hence both the marine-clay and river-clay areas contain deposits which are indicative of interrupted occupation. In fact, a large share of the Dutch soils consists of material deposited during historical time. Information obtained from studies on the age of the parent material and the horizontal and vertical distribution of layers of various ages is of great use to archaeologists. Traces of settlement found in the deeper sediments can sometimes be dated on the basis of this information, which may also provide clues about potential locations of traces of habitation.

In The Netherlands the usefulness of a soil survey to archaeologists has been recognized for a long time (Modderman, 1948a, b; Edelman, 1951), leading to a profitable co-operation between the two disciplines. This is illustrated in the present article by results of a soil survey (scale 1:10,000) made in Midden-Westfriesland (Du Burck and Dekker, in prep.). Although the findings apply to that particular area, the basic principles have a wider, more general application. Therefore, the results are communicated in this series of articles on the application of soil survey for non-agricultural purposes.

ORIGIN AND STRATIGRAPHY OF THE SOIL MATERIAL OF MIDDEN-WESTFRIESLAND

Midden-Westfriesland is part of The Netherlands coastal region (Fig.1) which was formed during the Holocene Period. Layers of marine sediments up to 15 m in thickness were deposited on Pleistocene materials after approx. 5500 B.C. Sedimentation took place in various stages. During transgressional stages clay and sand were deposited. During peri-

Fig.1. Position of area surveyed.

ods of regression, especially the later ones, peats were formed locally. These regressions permitted habitation by man.

In the upper 5 m of parent material in Midden-Westfriesland the following stages of sedimentation are distinguished and designated by The Netherlands Geological Survey:

Dunkirk III	(800–1400 A.D.)
Dunkirk IA	(700?–500?B.C.)
Dunkirk O	(1500–1200 B.C.)
Calais IVB	(2200–1500 B.C.)
Calais IVA	(2800–2200 B.C.)
Calais III	(before 2800 B.C.)

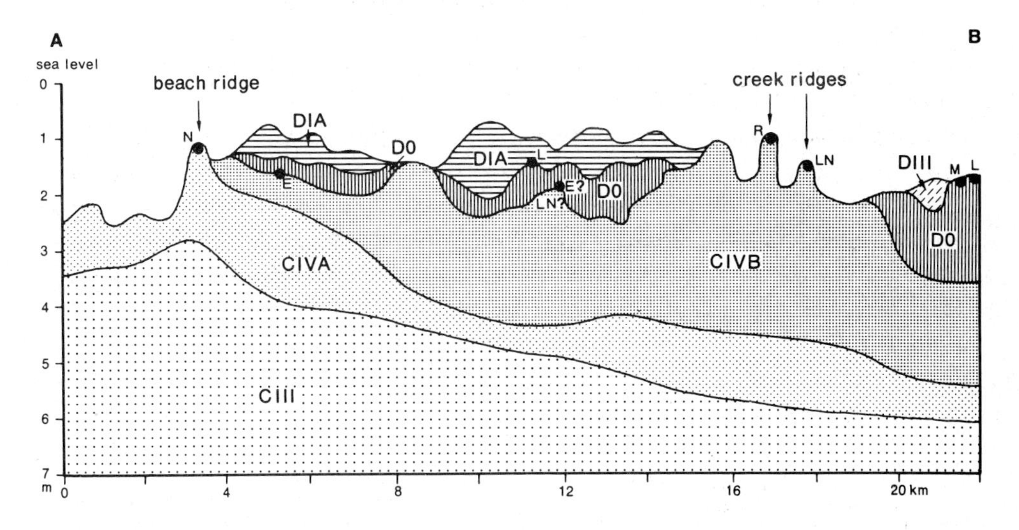

Marine deposits

DIII Dunkirk III (800-1400 A.D.)
DIA Dunkirk IA (700? - 500? B.C.)
DO Dunkirk 0 (1500-1200 B.C.)
CIVB Calais IVB (2200-1500 B.C.)
CIVA Calais IVA (2800-2200 B.C.)
CIII Calais III (>2800 B.C.)

Settlement traces (•):

R Roman Iron Age (circa 100-300 A.D.)
L Transition Late Bronze Age/Early Iron Age (800-500 B.C.)
M End of Middle Bronze Age (1200-1000 B.C.)
E Early Bronze Age: Barbed Wire Group (1700-1500 B.C.)
LN Late Neolithic: Bell Beaker Culture (2000-1700 B.C.)
N Late Neolithic: Vlaardingen Culture + Protruding Foot Beaker Culture (2500-2000 B.C.)

Fig. 2. Cross-section of the study area and location of a number of archaeological finds.

The Calais III deposit occurs only at depth and is not exposed at the surface. Calais IVA, Calais IVB and Dunkirk O lie partially at the surface and partially beneath younger deposits. The Dunkirk IA and III deposits are found only at the surface (Fig.2 and 3).

Between the deposits, thin layers of peat and layers of humose clay can be found in many places. Sometimes borings to a depth of 5 m have produced as many as four of these layers. The various deposits were dated by means of palynology and C^{14}-measurements of the peat and humic mineral layers. Due to lack of usable representatives of these layers the Dunkirk IA dating is still uncertain.

The cross-section (Fig.2) shows the vertical order of the various deposits. The position of the section is shown in Fig.3 which is based on the soil map. The units of Fig.3 have the same names as the deposits at the surface, because the main classification of soils in the area reflects the age and character of the parent material.

With the exception of a moderately coarse-textured beach ridge (M50: 210–420 μm), the parent material ranges from very fine sand (M50: 50–105 μm) to clay. Over a distance of 100 m the surface often shows differences in elevation of 2–3 m and soil conditions vary strongly, especially in texture and drainage. The irregular and uneven soil pattern was caused by the process of sedimentation. Marine erosion resulted in channels with many branches. In the channels of this highly irregular creek system, sand and loam were deposited, whereas clay was laid down at places where the current was negligible. When such an area was left dry, reversal of relief took place through settling of the clayey deposits. This resulted in a landscape of creek ridges and back swamps. Such a creek pattern is given in Fig.4. The soil surface differs from slightly below NAP* on the highest and largest ridges to approx. 3 m below NAP in the back swamps.

ARCHAEOLOGICAL FINDS IN THE LIGHT OF SOIL CONDITIONS

During the soil survey, potsherds lying on the surface were gathered from many parcels of arable land; occasionally sherds were brought up from the deeper layers by soil augers** In various parts of the area excavations had been conducted by archaeologists (Van Giffen, 1930, 1944 and 1961; Van Regteren Altena, 1962; Modderman, 1964; De Weerd, 1966; Van Regteren Altena and Bakker, 1968). The places where prehistoric and Roman Iron Age traces of occupation were found are shown in Fig.3: in addition to the ages of the artefacts, the legend differentiates between finds at the surface and those from deeper sediments.

In the Calais-IVA deposit remains from the Neolithic Period have been found in some places. Settlement traces of the Vlaardingen Culture (2500–2200 B.C.) and the Protruding Foot Beaker Culture (2300–2000 B.C.) have been discovered at the surface in a moderately coarse-textured beach ridge. In two other places artefacts of the Protruding Foot

*NAP = Normal Amsterdam Level, Ordnance Datum (more or less corresponding with the average sea level).

** The sherds have been identified by Drs. J.A. Bakker and Drs. R.W. Brandt of the Institute for Pre- en Protohistory, University of Amsterdam.

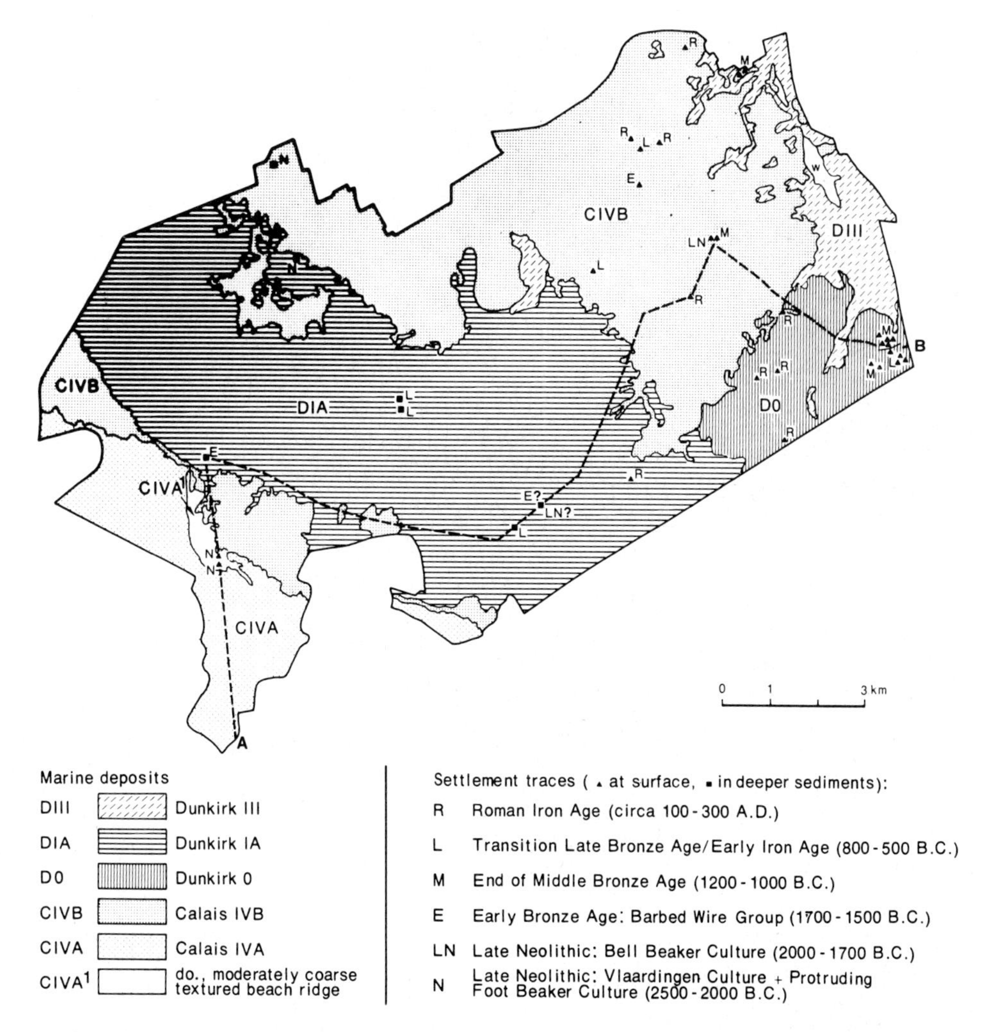

Fig.3. Distribution of deposits at the surface and of settlement traces in Midden-Westfriesland (see for *A-B* in Fig.2).

Beaker Culture have been recovered from the uppermost part of relatively high-lying sandy creek deposits. In one of these places a Calais-IVB layer of medium-textured material overlays the occupation level, in the other there is, in addition, a thin layer of Dunkirk IA.

In the Calais-IVB deposit settlement traces have been found from the Late Neolithic Period, the Early Bronze Age, the Middle Bronze Age, the transition Late Bronze Age/ Early Iron Age and from the Roman Iron Age. It is clear from Fig.4 that all settlement traces occur in the coarse and medium-textured soils of creek ridges. The Calais-IVB deposit is rich in lime so that during an excavation an almost 4000-year-old skeleton of an occupant of the Bell Beaker Culture was found completely intact (Fig.5).

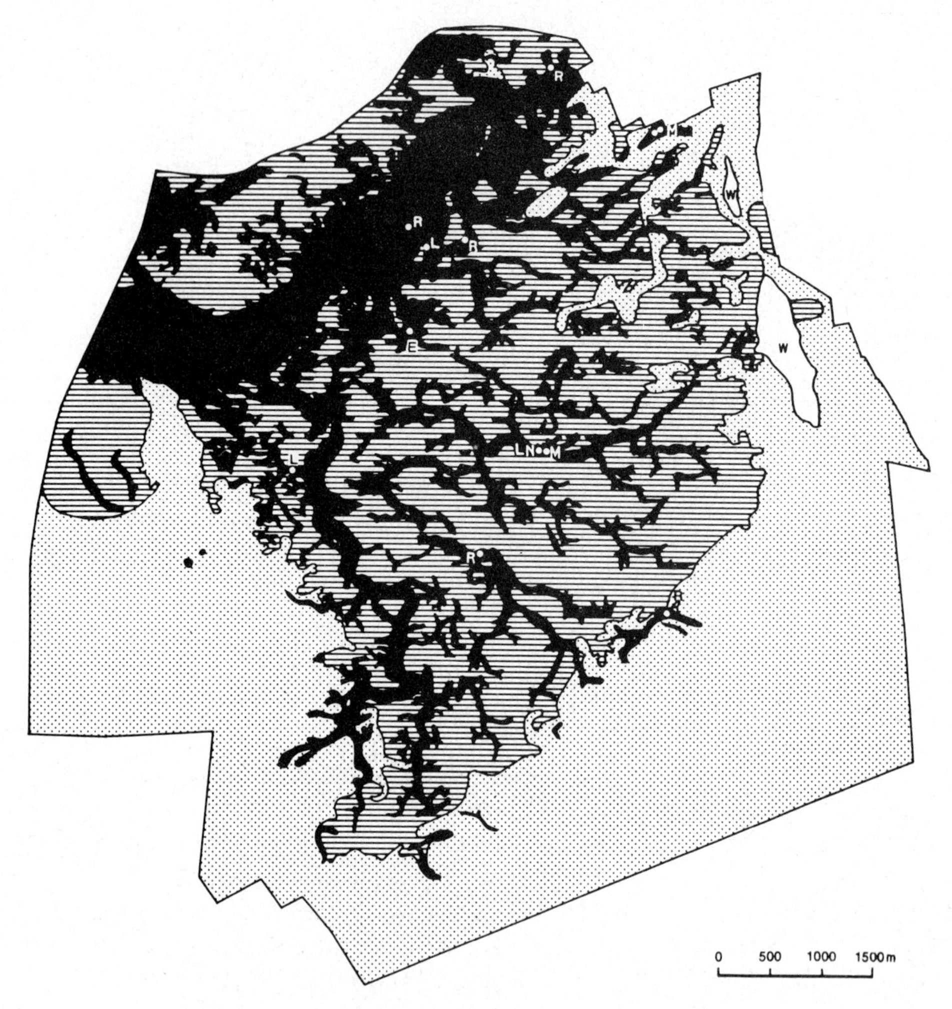

Settlement traces:

R Roman Iron Age (circa 100 - 300 A.D.)
L Transition Late Bronze Age/Early Iron Age (800 - 500 B.C.)
M End of Middle Bronze Age (1200 - 1000 B.C.)
E Early Bronze Age: Barbed Wire Group (1700 - 1500 B.C.)
LN Late Neolithic: Bell Beaker Culture (2000 - 1700 B.C.)

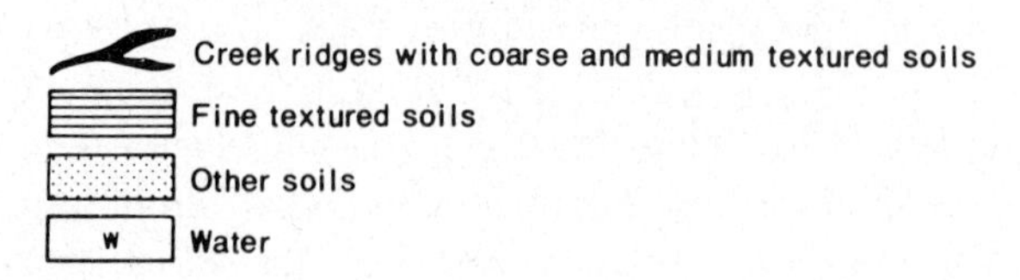

Fig.4. Detailed map of coarse and medium textured soils of creek ridges and fine textured soils developed in Calais IVB deposits, and distribution of settlement traces.

Fig.5. Flat grave (with traces of lateral shuttering) with virtually intact skeleton of an occupant of the Bell Beaker Culture (2000–1700 B.C.).
Excavation by Institute for Pre- en Protohistory, Amsterdam, 1963. Photograph F. Gijbels, IPP.

In the central and western part of the area, the Calais IVB is covered by Dunkirk 0 and Dunkirk IA deposits (see also Fig.2 and 3). During the soil survey archaeological artefacts were found at a depth of approx. 80 cm in the upper part of the Calais-IVB deposit in two places. The stratigraphic position of these finds is shown in Fig.2. The sherds at one place were found to be Early Bronze Age (Barbed Wire Group 1700–1500 B.C.), whereas at the other place they were not significant enough to warrant dating. On the basis of stratigraphy it may be assumed that the settlement traces date from the Late Neolithic Period or the Early Bronze Age. It is worth mentioning that at both places the finds lay on sandy creek deposits.

The Dunkirk-0 deposit crops out in the southeastern part of the area, consisting mainly of coarse and medium-textured material. The upper part lies at approx. 1.5 m below NAP. Soon after the end of sedimentation (approx. 1200 B.C.) the surface was occupied. This is apparent from the presence of five flat graves and a large number of barrows from the 12–11th century B.C. There was also human occupation during the Late Bronze Age/ Early Iron Age and during the Roman Iron Age. During the soil survey some settlement soils from the first-mentioned period were discovered. They are characterized by their slightly higher position, the deep dark colour and considerable thickness (often more than 50 cm) of the humose layer. This layer contains sherds and fragments of bone, and iron-phosphate stains were found in the subsoil.

In a large part of the area the Dunkirk-IA deposit overlies the Dunkirk 0 (Fig.2 and 3). Between these deposits there is often a thin humic layer at a depth of 0.5–1 m, in which artefacts from the Late Bronze Age or Early Iron Age have been recovered, always on top of sandy creek deposits.

It is remarkable that not any prehistoric sherd and only one piece of native pottery from the Roman Period have been found in the Dunkirk-IA deposit, though it is the surface-layer in a large part of the area and consists of high-lying coarse and medium-textured material.

Consistent with the age of the deposit is the absence of old artefacts in Dunkirk III.

Archaeological finds dating from the Middle Ages are scattered in large numbers in all soils of the area.

DISCUSSION AND CONCLUSIONS

There is a distinct relation between the age of deposits and settlement traces found in them. The oldest traces were recovered from the oldest deposits; the youngest deposits did not contain any settlement trace (see Fig.3). If a deposit crops out, archaeological finds may in principle be expected from various periods after the end of sedimentation.

The age of settlement traces, found in thin peat or humic layers beneath the surface, is closely related to the end of sedimentation of the underlying deposit (*terminus post quem*) and the onset of sedimentation of the deposit above it (*terminus ante quem*). This relation is clearly shown in Fig.2. The relation between the age of deposits and settlement traces is shown in Table I.

TABLE I

Settlement traces discovered and to be expected in deposits of various ages*

	Calais IVA 2800–2200 B.C.	Calais IVB 2200–1500 B.C.	Dunkirk 0 1500–1200 B.C.	Dunkirk IA 700?–500? B.C.	Dunkirk III 800–1400 A.D.
Roman Iron Age approx. 100–300 A.D.	/	X	X	X	–
Transition Late Bronze Age/Early Iron Age 800–500 B.C.	/	X	X ○	?	–
End of Middle Bronze Age 1200–1000 B.C.	/	X	X	–	–
Early Bronze Age: Barbed Wire Group 1700–1500 B.C.	/	X ○	–	–	–
Late Neolithic: Bell Beaker Culture 2000–1700 B.C.	/	X	–	–	–
Late Neolithic: Vlaardingen Culture + Protruding Foot Beaker Culture 2500–2000 B.C.	X ○	–	–	–	–

* Explanation of symbols:
X = settlement traces found at the surface
○ = settlement traces found beneath a younger deposit
/ = settlement traces not found, yet may be expected
– = no settlement traces found and not to be expected in view of age of deposit
? = not yet known

The coarse and medium-textured soils of creek ridges, well suited for a road network, were especially used as arable land, settlements and burial grounds during prehistoric times. A good example of this is shown in Fig.4.

The artefacts recovered during the soil survey have contributed much towards the knowledge of early human occupation. Settlement traces from the Early Bronze Age, Late Bronze Age/Early Iron Age and Roman Iron Age had not been found before in Midden-Westfriesland.

Until now archaeological investigations have been strongly dependent on discoveries made by accident. However, the number of places where remains are found may be enlarged by means of an archaeological survey. This can be effected quickly and efficiently by making use of a soil map. This is fast becoming normal procedure for archaeologists where regional soil maps are available. A large number of artefacts may be expected by systematic searching of the surfaces of creek ridges situated in arable land; there is very little chance of finding anything in fine-textured soils and in soils developed in young deposits.

The current archaeological explorations of the land-development area of Midden-Westfriesland, especially in road construction and in newly dug ditches, may result in settlement finds in the deeper layers. The knowledge of both development and density of human occupation may be extended and deepened by these explorations.

Knowledge of soil conditions is of crucial importance to a better understanding of archaeological finds. The more there is known of the soils, the better the basis will be for discussions on ancient occupation.

REFERENCES

Du Burck, P. and Dekker, L.W. (in preparation). *De Bodemgesteldheid van de Vier Noorder Koggen.* Stichting voor Bodemkartering, Wageningen.

De Weerd, M.D., 1966. Nederzettingssporen van de vroege klokbekercultuur bij Oostwoud (N.H.). *In het Voetspoor van A.E. van Giffen* (2nd ed.), Groningen, pp. 174–175.

Edelman, C.H., 1951. Archaeological results from soil surveys. *Boor Spade*, 4: 307–325 (in Dutch with a summary in English).

Modderman, P.J.R., 1948a.) Archaeology and soil science. *Boor Spade*, 1: 70–72 (in Dutch with a summary in English).

Modderman, P.J.R., 1948b. Archaeological aspects of the soil survey. *Boor Spade*, 2: 209–212 (in Dutch with a summary in English).

Modderman, P.J.R., 1964. Middle Bronze Age graves and settlement traces at Zwaagdijk, gemeente Wervershoof, Province of North Holland. *Proc. State Serv. Archaeol. Invest. Neth.*, 14: 27–36.

Van Giffen, A.E., 1930. *Die Bauart der Einzelgräber.* Mannus-Bibliothek Band 44 (208 pp.) and 45 (119 Tafel). Leipzig.

Van Giffen, A.E., 1944. Grafheuvels te Zwaagdijk, gem. Wervershoof, N.H. *West-Frieslands Oud Nieuw*, 17: 121–231.

Van Giffen, A.E., 1961. Settlement traces of the Early Bell Beaker Culture at Oostwoud (N.H.). *Helinium*, 1: 223–228.

Van Regteren Altena, J.F., 1962. Zandwerven. In: J.F. van Regteren Altena, J.A. Bakker, A.T. Clason, W. Glasbergen, W. Groenman-Van Waateringe and L.J. Pons (co-authors), *The Vlaardingen Culture. Helinium*, 2: 7–13.

Van Regteren Altena, H.H. and Bakker, J.A., 1968. Opgravingen bij Medemblik. *West-Frieslands Oud Nieuw*, 35: 201–210.